ALAS, POOR DARWIN

Hilary Rose, Visiting Research Professor at City University, London, is a feminist sociologist. Her most recent book is *Love, Power and Knowledge*. Steven Rose, a neuroscientist, is Professor of Biology at the Open University, and most recently author of *The Making of Memory* and *Lifelines: Biology, Freedom and Determinism*. They have worked and written together on issues in science and society for many years. They are currently joint Professors of Physic (genetics and society) at London's Gresham College and their jointly written and edited books include *Science and Society*, *The Radicalisation of Science* and *The Political Economy of Science*.

ALAS, POOR DARWIN

Arguments Against
Evolutionary Psychology

EDITED BY
Hilary Rose and Steven Rose

VINTAGE

Published by Vintage 2001

2 4 6 8 10 9 7 5 3 1

First published in Great Britain in 2000 by
Jonathan Cape

Vintage
Random House, 20 Vauxhall Bridge Road,
London SW1V 2SA

Random House Australia (Pty) Limited
20 Alfred Street, Milsons Point, Sydney
New South Wales 2061, Australia

Random House New Zealand Limited
18 Poland Road, Glenfield, Auckland 10,
New Zealand

Random House (Pty) Limited
Endulini, 5A Jubilee Road, Parktown 2193,
South Africa

The Random House Group Limited Reg. No. 954009
www.randomhouse.co.uk

A CIP catalogue record for this book
is available from the British Library

ISBN 0 09 928319 0

Printed and bound in Great Britain by
Bookmarque Ltd, Croydon, Surrey

Contents

1

Introduction

Hilary Rose and Steven Rose

Why is this book important? Because it brings together a multidisciplinary group of authors with the shared aim of challenging what we feel has become one of the most pervasive of present-day intellectual myths. Over the last ten years the number of books whose titles invoke Charles Darwin, the theorist of evolution by natural selection, has grown dramatically. 'Darwinian' and 'evolutionary' have become adjectives to attach to almost anything. Not only do we have evolutionary biology, medicine, psychology and psychiatry; there are evolutionary economics and evolutionary sociology. The term 'Darwinian' is employed to explain processes as seemingly varied as the origin of the universe, the expansion of companies on the Internet and the growth and competition of rival scientific theories. 'Darwinian methods' are supposed to underlie everything from computer technology to the processes of human thought. One philosopher, Daniel Dennett, has described Darwinism as a 'universal acid' that eats through everything it touches.

Among the disciplines rebranding themselves with the prefix 'evolutionary', the most influential has been evolutionary psychology. Evolutionary psychology, henceforward EP, is a particularly Anglo-American phenomenon.[1] It claims to explain all aspects of human behaviour, and thence culture and society, on the basis of universal features of human nature that found their final evolutionary form during the infancy of our species some 100–600,000 years ago. Thus, for EP, what its protagonists describe as the 'architecture of the human mind' which evolved during the Pleistocene is fixed, and insufficient time has elapsed for any significant subsequent change. In this architecture there have been no major repairs, no extensions, no refurbishments, indeed nothing to suggest that micro or macro contextual changes since prehistory have been accompanied by evolutionary adaptation. The extreme nature of this claim, granted the huge

1

changes produced by artificial selection by humans among domesticated animals – cattle, dogs and even Darwin's own favourites, pigeons – in only a few generations, is worth pondering. Indeed, unaided natural selection amongst the finches in Darwin's own islands, the Galapagos, studied over several decades by the Grants[2] is enough to produce significant changes in the birds' beaks and feeding habits in response to climate change. If for birds and beasts, why not humans?

To evolutionary psychologists, everything from children's alleged dislike of spinach to our supposed universal preferences for scenery featuring grassland and water derives from this mythic human origin in the African savannah. And of course there are more serious claims, such as those legitimising men's 'philandering' and women's 'coyness', our capacity to detect cheaters, to favour our genetic kin, to be aggressive. Evolutionary psychologists claim to have identified and explained all these as biological adaptations – that is, behaviours that have been selected during human evolution to assist in survival and hence the propagation of our ancestors' genes. The main players in this new genre are the psychologists Leda Cosmides and John Tooby, Margo Wilson and Martin Daly, Steven Pinker, and the several disseminators of their ideas from the science writers Robert Wright and Matt Ridley to Helena Cronin, organiser of the London-based Darwin Seminar.

Perhaps the nadir of evolutionary psychology's speculative fantasies was reached earlier this year with the publication of *A Natural History of Rape: Biological Bases of Sexual Coercion*, by Randy Thornhill and Craig Palmer. In characteristic EP style, Thornhill and Palmer argue that rape is an adaptive strategy by which otherwise sexually unsuccessful men propagate their genes by mating with fertile women. To make this claim they draw extensively on examples of forced sex among animals, which they insist on categorising as 'rape'. Yet as long ago as the 1980s the leading journals in the field of animal behaviour rejected this type of sociobiological strategy which anthropomorphises animal behaviour. Specifically, using the term 'rape' to refer to forced sex by mallard ducks or scorpionflies (Thornhill's animal of study) was ruled out, as it is not a helpful concept in the non-human context because it conflates conspicuous differences between human and other animals' practices of forced sex. Above all forced sex among animals always takes place with fertile females – hence the reproductive potential. As those women's groups, lawyers and feminist criminologists who have confronted rape over the last three decades have documented, victims of rape are often either too young or too old to be fertile. The universalistic explanation offered by Thornhill and Palmer simply fails to address the evidence. Instead they insult women, victims and non-victims alike, by suggesting, for example, that a tight blouse is in itself an automatic invitation to sex.[3] They insist on distal (in their slightly archaic language, 'ultimate') explanations when proximate ones are so much more explanatory (see Steven Rose's chapter). Further, given the difficulties of securing convictions, and the immense guilt which still surrounds rape victims so that tragically they feel they have brought rape on themselves, the measurements of the incidence of rape are extremely frail. Despite their protestations that they want to help women, the version of evolutionary psychology offered by Thornhill and Palmer is offensive both to women and also to the project of building a culture which rejects rape.

To an uninitiated eye, evolutionary psychology, which seems to have got into the cultural drinking water in both the USA and the UK, may seem little different from old-style sociobiology, whose exponents, E. O. Wilson, Richard Dawkins, Robert Trivers and David Buss, still find a place in EP's pantheon of intellectual heroes. And indeed there are major continuities, but, as both we and the founders of EP agree, there are also important differences. It is the argument of the authors of this book that the claims of EP in the fields of biology, psychology, anthropology, sociology, cultural studies and philosophy are for the most part not merely mistaken, but culturally pernicious. Further they claim that their new view of human nature should inform the making of social and public policy. Thus the new science has a directly political dimension, although its protagonists vary in their advocacy of both the direction and the speed of implementation.

But to understand the appearance and significance of EP, it is necessary to set this new discipline in a historical and social context. The last decades of the twentieth century have been a period of almost unprecedented social, economic and cultural turbulence. Many of the old seeming certainties and indeed hopes for a more socially just future have crumbled. The collapse of communism, the end of the Cold War and the bloody regional nationalist and ethnic struggles that it has unleashed, the weakening of the welfare state and growing fears of ecological catastrophe have been matched by a shaken belief in the inevitability of progress. The unquestioned benevolence of scientific and technological rationality can no longer be assumed.[4]

In this climate the search for new apparent certainties, something to cling on to, has become urgent but extraordinarily diverse. There has been a resurgence of fundamentalist religions, Islamic, Jewish and Christian, with their enthusiasm for militancy and their beliefs in creationism. Alongside these have appeared the New Age movements, which explicitly reject scientific modernism and seek a new spiritual relationship between people and nature. Spectacular mobilisations in defence of green nature, of trees, land and water, have largely replaced the great industrial working-class struggles. But that old resistance to an unconstrained capital has not gone away; it has reappeared in even more profound and worldwide form as a battle for the very survival of the human ecosystem against the globalised markets. The iconic moment of this new resistance was surely the dramatic battle in Seattle against the World Trade Organisation in the final months of the twentieth century.

While the influential but still minority environmentalist movement confronts technoscientific rationality, the dominant and increasingly secular technoscientific culture within the UK and USA[5] becomes increasingly geneticised. As the human genome is mapped and sequenced we also learn that DNA is the common denominator of all life forms. That we share 98 per cent of our DNA with chimps and 35 per cent with daffodils potentially changes not only how we think about ourselves but also our connection with other life forms.[6] Part of and parallel with this geneticisation of culture are the twin technologies of informatics and genetic engineering, unquestionably the technosciences set to dominate the twenty-first century.

Within this framework new forms of biological determinism have arisen. Not of course the old Social Darwinism, which provided a biologised defence of the prevailing social hierarchies of Victorian society, but modern and seemingly more sophisticated forms, powered by this vast expansion of biological, and particularly genetic, knowledge and technology. This new determinism takes two apparently antithetical forms. On the one hand, it claims, our biology is our destiny, written in our genes by the shaping forces of human evolution through natural selection and random mutation. This biological fatalism is opposed by Promethean claims that biotechnology, in the form of genetic engineering, can manipulate our genes in such a way as to rescue us from the worst of our fates. It offers to eliminate illness, prolong life, grant our children enhanced intelligence and better looks – a cornucopia of technological goodies undreamed of even in the science fiction of prior generations.

These contrasting claims have been fanned by a host of popular books, television programmes and films: *Genes Are Us* meets *Designer Babies*. It not just the media which projects a future of genetically engineered people, as in the 1990s' movie *Gattaca*; molecular biologists themselves get into the futurist game. Princeton geneticist Lee Silver's bestselling book *Remaking Eden* shares essentially the same plot as the Hollywood science-fiction movie. Unsurprisingly such media and scientific brouhaha, amplifying fantastic fears and hopes, is reflected in popular opinion surveys in which people say that they would indeed like to select special attributes in their children, or be cloned, or whatever. More careful social research, such as that carried out by the Wellcome Trust on human cloning,[7] which gives its subjects space for thoughtful reflection, shows much greater scepticism and caution.

That second strand, of biological fatalism, however, runs alongside and interacts with the Promethean claims of these promised genetic technologies. This strand is above all part of a struggle about how we should conceive of human nature and human society and culture. Of course, the claims of biology-as-destiny are as old as history itself, a continuing and powerful cultural strand in an old narrative. Typically, such accounts find reflected within nature the dominant social hierarchies of their time. But the deforming fingers of the culture they produce persist long after both the knowledge-producers and indeed even the social hierarchies their biology celebrated have turned into dust. Thus the feminist movement of the 1970s found it necessary, not only to confront contemporary biologised theories of patriarchy, but to go back to Aristotle as a foundational and deeply patriarchal figure within Western thought to confront his theory of human nature. Feminists saw it as crucial to expose the inadequacy of his biologised account of male superiority in order to challenge the trouble this was still causing when trying to understand late twentieth-century gender relations. As well as justifying the lower place of women in ancient Greek society, Aristotle also provided a biological justification of slavery. He divided the population of slaves into 'natural' and 'unnatural' – the latter being brave warriors captured in battle, the former mere brutes fit for menial labour. To free natural slaves was thus to go against nature; to free unnatural slaves a social recognition of their high natural status.

The predestinationist claims of the nineteenth and twentieth centuries divided the human world along new faultlines which predestinationist biology faithfully began to mirror in the concepts of race, class, gender and ethnicity. The biology of the French race-theorist Comte Joseph de Gobineau, the British gender-theorist Herbert Spencer and class theorists Thomas Malthus and Francis Galton was transmuted into policy early in the twentieth century by the US eugenicists Charles Davenport and Henry Goddard and the Nazi geneticists Fritz Lenz and Erwin Baur. These are familiar names in the list of those drawing on biologised narrative to justify forms of social domination, from discrimination and exclusion to genocide.

The Nazi genocide of the gypsies and the Jews generated such worldwide revulsion that many felt that this of itself would terminate such evil racism. Certainly the widespread support during the 1930s for eugenics by left and liberal intellectuals, feminists, geneticists and welfare reformers faded in the face of the horror of the death camps. Thus to dissociate science from racism became a crucial cultural objective at the end of the Second World War. To secure this goal, the newly formed UNESCO commissioned a statement on race by a group composed primarily of US cultural anthropologists inspired by Franz Boas.[8] The statement, published in 1950, stressed the unity of human peoples, fundamentally rejecting the idea of race as a scientific concept. Subsequently geneticists such as Theodosius Dobzhansky and Herman Muller partially reformulated the statement, but the argument that the concept of race was not science but psuedo-science remained constant. The problem with this basically well-intentioned move was that, while it is true that race may have no useful purchase in understanding the diversity of human biology, race and racism were and are virulently alive in cultural and political reality. What the UNESCO statement had done was first to construct a seemingly impermeable barrier between the social and the biological sciences, and second to assume that the latter, as a natural science, had the greater authority. This belief in the superior power of biology to name and therefore own phenomena persists, so that we find the British geneticist Steve Jones (professor in the Galton laboratory of genetics at University College London) claiming in the 1990s to the journalist Marek Kohn that race would not return to science. When Kohn objected that race was thriving in psychology, Jones, echoing this assumption of the superiority of the natural sciences, replied that psychology was not science.[9]

Thus, despite UNESCO, biologically determinist ideas lived on. During the 1960s they took the form of pop-ethology books stressing the evolutionarily determined innateness of human aggression (as in the writings of the ex-Nazi ethologist Konrad Lorenz or the American science writer Robert Ardrey). The claim of black/white racial differences in intelligence resurfaced in 1969 with the writings of Arthur Jensen in the USA and Hans Eysenck in the UK – forerunners of Richard Herrnstein and Charles Murray's *The Bell Curve* in the 1990s. By the early 1970s sociologist Steven Goldberg was arguing for *The Inevitability of Patriarchy* on the basis of hormonal differences between men and women. Richard Dawkins's *The Selfish Gene* and E.O. Wilson's *Sociobiology: The New Synthesis* sought to transcend these early works (Hilary Rose's chapter maps this process in more detail) by providing what seemed like an all-embracing theoretical framework. Within this the human condition could be seen as an epiphenomenon of evolution by natural selection based on the 'drive' by individual genes to reproduce copies of themselves. The pretensions of sociobiology – at least insofar as they reflected on humans as opposed to other species – were soon under heavy criticism in such books as *Biology as a Social Weapon* (the Science for the People Editorial Collective), *The Uses and Abuses of Biology* (Marshall Sahlins), *Not In Our Genes* (Steven Rose, Richard Lewontin and Leo Kamin) and the *Genes and Gender* series (edited by Ethel Tobach and Betty Rosoff).

Despite the political implications for both the women's and black civil rights movements, however, the cultural struggles around the central concepts and claims of sociobiology were primarily waged between biologists, although they were followed by a much wider public. The attacks came from left, liberal and feminist biologists together with a handful of non-biologists, notably the US anthropologists Marshall Sahlins and Ashley Montagu, who had drafted the original 1950 UNESCO statement. Slightly later the attack was joined by the elegant critique of the US philosopher and historian of science Philip Kitcher (*Vaulting Ambition*). By the late 1980s the conflict settled down into a stalemate. US school curricula, for example, found it necessary to include both sociobiology and its critics. The criticism of sociobiology as updated Social Darwinism, with all that that entailed for support of elite white male social dominance, stuck. In response sociobiologists themselves muted their claims that their discipline was primarily concerned to explain human nature, and instead refocused their discipline as a branch of animal behaviour – behavioural ecology.

It was at this point that a new group of players entered the game. Psychologists, sometimes with a past in studying animal rather than human behaviour, together with cognitive psychologists and biological anthropologists, were all attracted to the view that evolutionary insights provided a powerful new tool for transforming the social sciences into soundly biologically based disciplines. Drawing on the reductionist claims of the earlier sociobiologists, both individual human minds and, even more controversially, the complex and multifarious forms of human society, were to be reduced to the – admittedly mediated – workings out of genetic and evolutionary imperatives. These claims were fanned by popularised accounts and found their way into day-to-day broadsheet newspaper discussions and TV programmes. Unlike sociobiology, they have been included in the teaching syllabuses of university courses in many areas of the social sciences and humanities without critical comment.

Thus while it is still the case, for example, that sociology is mostly immune to the attractions of EP, some sociologists, not least the influential British theorist W. G. Runciman, have been attracted to evolutionary theory. His work has been sympathetically discussed in the major theoretical journal *New Left Review*, which indicates something of the complicated cultural and political reaction to evolutionary theory. Elsewhere the political agenda of EP is transparently part of a right-wing libertarian attack on collectivity, above all the welfare state. Thus, as Hilary Rose's chapter explains, it shows itself to be recruitable around almost any political agenda – except that of racism, which is ruled out by its insistence on the unity of the human species.

What we wrote above about sociobiology and the difference between the reaction of biologists and, for example, sociologists was also true for the Roses. As a biologist, he critically engaged with sociobiology. As a sociologist, she welcomed the criticism, but felt that it was the task of biologists to take the argument on from within their own canon. There was only one occasion, during the 1980s, when we felt it necessary to write a joint essay, in the journal *Critical Social Policy*, showing how sociobiology was culturally underpinning the Thatcherite attack on the welfare state. Our response to EP has been very different. The sociologist and the biologist separately felt that EP was making insupportable assertions which touched our own distinctive fields. For the biologist who had researched brain processes for nearly forty years, the flashpoint was the argument that the mind is innately structured into modules which work like a Swiss army knife.[10] For the sociologist and feminist it was the claim that parental love is reducible to genetics. Because EP theorists attack the social sciences as a preliminary move before constructing their new approach, they directly challenge social science in a way sociobiology did not. It is interesting that in his latest book, *Consilience*, the sociobiologist E.O. Wilson borrows from EP's critique of the social sciences to shore up his own position. Thus unlike the foundational texts of sociobiology, revisionist sociobiology along with the new EP demands a reply from within the canon of the social sciences.

This sense that it was no longer possible to be a spectator while radical biologists attacked the old monolithic biological determinism of sociobiology was widely shared. Cultural theorists, philosophers, anthropologists, developmental psychologists, ethologists and molecular biologists were drawn into criticising claims which affected their fields and hence general culture. Thus in a moment of exasperation with some of EP's more fantastic claims, the ethologist Pat Bateson complained that it had set back evolutionary theory by ten years. A colleague teaching family studies to social work students saw the plausibility of the EP theory of stepfather infanticide as troublingly attractive to those students who wanted easy answers to complex problems.

Hence, to return to our opening sentence, our reason for claiming the importance of this book. Its history is worth telling. We conceived of it early in 1998. We recognised that, by contrast with the case of sociobiology, this time the critiques of EP must of necessity come from many disciplines. But we wanted to ensure that as far as possible those who spoke in these many voices had the opportunity to discuss their ideas with one another and mutually hone them before presenting them in book form. Such an opportunity was provided by the postmodernist architectural critic Charles Jencks, who hosted a three-day seminar in September 1998 in the Keswicks' family house, Portrack, in Scotland. Portrack has the unusual charm of a DNA garden of the senses designed by Charles together with his late wife, the landscape gardener Maggie Keswick. Fifteen of the potential contributors met to discuss the pre-circulated drafts that formed the basis for most of the chapters in this book. We have undertaken the task of integrating and editing the papers resulting from these immensely exciting and stimulating few days.

The contributors come to the discussion of evolutionary psychology from many different points of view and do not speak with a single voice. Some of the resulting chapters are primarily critical of the claims of some of EP's most prominent writers, others are concerned with developing alternative explanatory perspectives – in biology, anthropology, developmental psychology and social theory. Both the criticisms and the alternatives are important. It's an old academic adage, but nonetheless true, that bad theory can never be driven out solely by criticism. A better alternative has to be offered. Critics of EP such as the science journalist John Horgan[11] and the religious affairs commentator Andrew Brown[12] have usefully interrupted the self-referential flow of the EP narrative, but replacing a flawed narrative has ultimately to come from within the disciplines themselves.

The book is organised as follows. In the first chapter, the sociologist Dorothy Nelkin reflects on the religious language and metaphor with which so much evolutionary psychology, particularly in its US variant, is suffused. EP, she shows, seeks to take its place within and alongside a form of scientific Christianity. Chapter 2 was written by Charles Jencks in the light of his encounter with E. O. Wilson – the man whom the novelist Tom Wolfe decribed as the new Darwin – at a conference in Boston. Here Wilson developed his argument for *Consilience* – the unity of the sciences and the subordination of art to human evolutionary imperatives – a claim which Jencks demolishes in a stylishly ironic essay. Chapter 3 is by the molecular biologist Gabriel Dover. His target, as indicated by the chapter's title, is Richard Dawkins's slogan, 'the selfish gene'. Dover's own research programme has led him to a different and more complex view of the nature of genetic mechanisms in evolution, notably the concept of adoptation, and his chapter serves both as a critique of Dawkinsian genetics and an explication of the alternative view.

One of the more bizarre extensions of selfish genery in recent years has been the way in which a number of EP converts have taken up an almost off-the-cuff suggestion by Dawkins that there might exist units of culture, analogous to genes, which could similarly replicate and mutate inside individual people. Our brains would in this sense become the lumbering robots required for memic transmission just as our bodies are, for Dawkins, for our genes. The philosopher Mary Midgley analyses this claim in her chapter. Neither thought nor culture, she insists, are granular or can be decomposed into memic units. Unlike genes, memes are content-free concepts. Chapter 5 sees the evolutionary theorist and writer Stephen Jay Gould confront philosopher David Dennett's simplistic understanding of Darwinian natural selection theory, in a characteristically robust essay which appeared in an earlier version in the *New York Review of Books*. Where Dennett sees Darwinism as a universal acid, Gould emphasises complexity, the plurality of evolutionary mechanisms and the naïveté of monolithic adaptationist explanations.

In Chapter 7 Hilary Rose tackles head on the claims of Tooby and Cosmides that an entity they call the 'standard social science model' is central to sociology and anthropology, and that it needs to be replaced by a model subordinate to evolutionary psychology's own perspectives. She traces the roots of EP from Malthus, Darwin and nineteenth-century Social Darwinism to the present day and discusses its relationship to both past and present Anglo-American neoliberalism. She contrasts the scientifically satisfying explanation of suicide by the nineteenth-century sociologist Durkheim (attacked by EP for defending the autonomy of the social) with Daly and Wilson's inadequate evolutionary explanation of infanticide.

One of the bestselling EP books of recent years has been Steven Pinker's *How the Mind Works*, billed on its UK dust jacket as the best book ever written about the mind. Cultural theorist Barbara Herrnstein Smith subjects Pinker's claims to close critical scrutiny in Chapter 7. She focuses on two aspects of evolutionary psychology which are abundantly present in Pinker's writing: first, the distinctive, pre-emptive claims issued by EP's practitioners and promoters and, second, the peculiarly sewn-up image of the mind that emerges from these accounts. She contrasts the premature triumphalism of EP, which declares victory for its team, with the rich developments taking place within contemporary cognitive science. Her chapter is followed by developmental child psychologist Annette Karmiloff-Smith's critique of nativism – the belief in innate characteristics – which exactly complements Herrnstein Smith's general point by empirically demolishing one of Pinker's central theoretical claims, that of the modularity of mind (the notorious Swiss army knife analogy). Pinker argues that mental processes are not only modular but innate, so that children are born, for instance, with a language acquisition device (LAD). Karmiloff-Smith's work with children suffering from Williams syndrome, who have characteristic language deficits, shows that the evidence cannot support the nativist LAD hypothesis.

Chapter 9, by Pat Bateson, retraces the old nature/nurture debate and the concept of instinct, with its roots in ethological theory and the writings of Konrad Lorenz. The developmental perspective which Bateson shares with Karmiloff-Smith transcends these old dichotomies. For Bateson living creatures are not usefully understood as partitioned into fixed and discrete components provided by genes and environment, or behaviour as partitioned between learned and instinctive. Process is central within any account of development and thence evolution.

Next, the feminist biologist Anne Fausto-Sterling crisply surveys the androcentric history of so much evolutionary writing, beginning with Darwin's initial failure to consider the evolution of women at all, to the recent claims of EP advocates David Buss and Robert Wright. These claim to fill the gap in evolutionary theorising, but continue to rehearse the nineteenth-century story of the natural persistence of female coynesss (despite the animal studies by Sarah Hrdy and other contemporary primatologists) and male ambition. Fausto-Sterling sharply contrasts the sophistication of the animal studies of mating with the methodological weaknesses of the allegedly parallel studies of human mating practices. Because biology-as-destiny has historically been associated with attempts to define some human Other as a lesser life and therefore to be stigmatised, excluded or worse, it is appropriate that disability sociologist Tom Shakespeare and his colleague Mark Erickson follow. Disability theorists, like many feminists, want to move beyond the either/or of the 'two cultures' model. Disability, like a baby, is at the same moment both social and biological, not one or the other.

With sociologist, amateur naturalist and Darwin scholar Ted Benton, we return to Darwin and to Alfred Russel Wallace. He reminds us of their simultaneous discovery of natural selection, but their separation over both human evolution and Darwin's androcentric focus. He then turns to a critical and detailed analysis of the attempt to introduce evolutionary theory into sociology by W. G. Runciman in his book *The Social Animal.* He shares Mary Midgley's contempt for memology, and Hilary Rose's insistence on the autonomy of the social, against the colonising pretensions of EP.

The anthropologist Tim Ingold, like Shakespeare and Erickson, Fausto-Sterling and Bateson, argues against either/or dichotomous thinking. Beginning with an account of several anthropological, cultural and psychological theorists who attempt, but, he argues, fail to transcend reductionism, he takes as his paradigm case that of walking. Can 'walking' be considered as an evolved, innately developing skill independent of the context in which individuals learn? Ingold's answer is emphatically no: we do not walk in the abstract, but only in a myriad concrete ways shaped simultaneously by our physical environment, our biology and our cultural milieu. In the last chapter Steven Rose seeks to escape the chains placed on the understanding of living processes by either the one-dimensionality and static character of DNA-based thinking about developmental and evolutionary processes or by that of the innate modular architecture of the mind conceived as an information-processing device. Minds, he argues, deal not with information but with meaning and living organisms must be understood not as reducible to their genes but as following a lifeline trajectory, simultaneously product and process, being and becoming.

Finally, we would like to make some acknowledgements. In a competitive academic world, it has been a delight to work with contributors none of whom attempted to impose their special disciplinary understandings on others, but who were prepared to listen and to collaborate in creating this book as, we trust, something more than the sum of its parts. This process was aided by Charles Jencks's generous hospitality at Portrack and by the input of others present at the seminar, but who are not represented in this book: Cambridge (US) molecular biologist Jeremy Ahouse, MIT psychologist Bob Berwick and Birkbeck College historian of science Dorothy Porter. Many friends and colleagues have provided helpful comments on the various chapter drafts. The continuous support of the book's agent Kay McCauley and editors Will Sulkin and Jörg Hensgen at Cape in London and Patty Gift at Crown in New York has been essential. Liz Cowen's meticulous copy-editing has made our task much easier. We hope that this sense of collaboration is reflected as much in the response of those who read this book as it has been for us as editors.

January 2000

Notes and References

1 Other European countries, notably France, have been less overwhelmed by Darwinian evolutionary theory.

2 Jonathan Weiner, *The Beak of the Finch* (London, Vintage, 1994).

3 Randy Thornhill and Craig Palmer, *A Natural History of Rape: Biological Bases of Sexual Coercion* (Cambridge, MA, MIT Press, 2000), p. 179.

4 It is interesting in this context that the theme of the Expo 2000 festival in Hanover in Germany is the need for the twenty-first century to clear up the scientific and technological mess left by the twentieth.

5 Though, given the levels of religiosity within the USA and the need for both politicians and scientists to handle that dimension cautiously, it is sometimes difficult to see the deep secularism of US science, especially the life sciences. However, a recent poll published in the US journal *Science* showed that a great majority of life scientists are now non-believers. Physicists are less hostile to religion – think, for instance, of such prominent writers as Paul Davies or Stephen Hawking.

6 Jonathan Marks, 'The Human Genome in Evolutionary Context: 98% Chimpanzee and 35% Daffodil', Symposium no. 124: Anthropology in the Age of Genetics, Wenner Gren Foundation, Teresopolis, Brazil, 11–19 June 1999.

7 Wellcome Trust Medicine and Society Programme, *Public Perceptions on Human Cloning* (London, Wellcome Trust, 1998).

8 The rapporteur for the group was the anthropologist Ashley Montagu, who was to engage in many of the later confrontations with, for instance, sociobiology and died only in 1999.

9 Marek Kohn, *The Race Gallery: The Return of Racial Science* (London, Cape, 1995), p. 7.

10 The EP authors most publicly associated with these propositions, though by no means the only ones, are Daly and Wilson, and Pinker.

11 John Horgan, *The Undiscovered Mind: How the Brain Defies Explanation* (London, Weidenfeld & Nicolson, 1999).

12 Andrew Brown, *The Darwin Wars: How Stupid Genes Became Selfish Gods* (London, Simon & Schuster, 1999).

2

Less Selfish than Sacred?
Genes and the Religious Impulse
in Evolutionary Psychology

Dorothy Nelkin

When Harvard University entomologist Edward O. Wilson first learned about evolution, he experienced, in his words, an 'epiphany'. He describes the experience: 'Suddenly – that is not too strong a word – I saw the world in a wholly new way . . . A tumbler fell somewhere in my mind, and a door opened to a new world. I was enthralled, couldn't stop thinking about the implications evolution has . . . for just about everything.'[1]

Wilson, who was raised as a southern Baptist, believes in the power of revelation. Though he drifted away from the Church, he maintained his religious feeling. 'Perhaps science is a continuation on new and better tested ground to attain the same end. If so, then, in that sense science is religion liberated and writ large.'[2]

Religion has been defined as a belief system that includes the idea of the existence of 'an eternal principle . . . that has created the world, that governs it, that controls its destinies or that intervenes in the natural course of its history'.[3] Believers understand this eternal principle – whether a God or a powerful idea – to be the key to all knowledge, the explanation of history, and the guide to the conduct of everyday behaviour.

According to a statement by the American Association for the Advancement of Science (AAAS), the differences between science and religion have to do with the kind of questions asked: 'Science is about causes, religion about meaning. Science deals with how things happen in nature, religion with why there is anything rather than nothing. Science answers specific questions about the workings of nature, religion addresses the ultimate ground of nature.'[4]

Yet scientists who call themselves evolutionary psychologists, including

14

those from the related disciplines of sociobiology and behavioural genetics, are addressing questions about meaning, about why things happen, about the ultimate ground of nature. Their explanations are based on the principle that human nature and human behaviour are governed by the evolutionary process of natural selection. According to this principle, people behave in ways that confer the greatest 'Darwinian fitness' for their offspring, that is, for the perpetuation of their genes.

Edward O. Wilson developed the all-encompassing dimensions of this principle in several books, including *Sociobiology* (1975), *On Human Nature* (1978) and *Consilience: The Unity of Knowledge* (1998). He claims that individual and cultural practices, including kin selection, parental investment, mating strategy, status seeking, territorial expansion and defence, and contractual agreements are all determined by the impulse to confer Darwinian advantage to the genes.[5] The eternal principle of natural selection, he believes, shapes our behaviour, moral impulses, human relationships and cultural norms.

He and other scientists have promoted this model of human nature in popular books and magazines with missionary fervour, aiming to convert the unenlightened. So ardent are their efforts, it is almost as if they aspire to assure the Darwinian fitness of the theory – to assure its survival in the world of cosmic ideas. Their claims, their language and their style have striking religious overtones.

In this chapter I do not attempt to evaluate the scientific validity of the ideas promulgated by evolutionary psychologists or sociobiologists. Others, not least in this volume, have done this well. Rather, I examine various ways in which these theories and the style of promoting them are motivated by a religious impulse. To be sure, their theories do not rely on a God or Divine Mover. Some sociobiologists like Richard Dawkins pride themselves on being materialist, reductionist and overtly anti-religious. But they offer theories proclaiming the evolutionary basis of human behaviour as explanations for virtually everything and as the basis for the unification of knowledge. Scientists promoting genetic explanations use a language replete with religious metaphors and concepts such as immortality and essentialism – indeed, the gene appears as a kind of sacred 'soul'. And as missionaries bringing truth to the unenlightened, they claim their theories are guides to moral action and policy agendas. They are, I argue, part of a current cultural move to blur the boundaries between science and religion.

A Theory of Everything

Biologists have long sought to unify knowledge through the elucidation of the fundamental properties of life.[6] In the 1930s in Britain and the United States, this effort took the form of the 'evolutionary synthesis', which seemed to reconcile Darwinism and Mendelism – selection and genetics – theories that were apparently contradictory. The architects of the synthesis promoted the idea that biological change through time – that is, evolution – could serve as the intellectual centrepiece for the study of life. In the 1950s the rise of molecular biology promised to explain life at its most fundamental physico-chemical level, the double helix of DNA.

Sociobiology and evolutionary psychology are but the latest efforts to develop a unifying theory that will explain the meaning of 'Life itself'.[7] In 1975 Wilson announced a 'new synthesis' that drew on both evolutionary biology and molecular biology to explain the human social order in biological terms. In subsequent years DNA, the so-called 'secret of life', became the most important entity in the search for an essential biological principle – reflected in the international efforts to map the human genome.

Molecular biologists have focused their work more on genetic diseases than on behaviour, but they are also exploring the genetic bases of mental illness, obesity and homosexuality, and some are pursuing genes that might lead to a propensity to violence. For the most part, however, they have left the complex and controversial terrain of human behaviour to psychologists who draw inferences about the heritability of behaviour from studies of identical twins,[8] or to sociobiologists and evolutionary psychologists who are developing theoretical arguments about the influence of natural selection on the human condition.

The theory of natural selection, they claim, explains why individuals engage in such complex behaviours as love, jealousy, risk taking, infidelity, rape, status seeking, violence and addiction. The desire for evolutionary fitness also lies at the root of cultural differences in gender distinctions and social relationships; and it defines our concepts of good and evil. Natural selection to evolutionary psychologists is a 'theory of everything', an eternal principle that explains why we behave the way we do and what makes us what we are; it defines the very meaning of human existence.

Though concerned about genes, evolutionary psychologists are no longer addressing the old debate about the relative influence of nature or nurture on human behaviour: they are firmly convinced of the biological basis of human nature and culture as well. They are rather seeking universal explanations – the cosmic truth that underlies life, death, culture and faith. This truth lies in natural selection as 'the consistent

16

guiding force'.[9] The need to maximise 'evolutionary fitness' governs the world, controls destiny, intervenes in history and guides the conduct of human behaviour.

Reviewing the field of evolutionary psychology, journalist Robert Wright's *The Moral Animal* – revealingly subtitled *Why We Are the Way We Are* – concludes that all our behaviour reflects the need to maximise genetic inheritance.[10] (See Anne Fausto-Sterling's chapter.) Robin Baker in *The Sperm Wars* offers evolutionary explanations of human sexuality: all sexual behaviour is driven by the need to pass on genes and sperm compete to maximise the genetic potential of offspring.[11] Richard Dawkins in one of the earliest books on sociobiology, *The Selfish Gene*, reduces people to the status of 'robot vehicles' programmed to perpetuate genes.[12] Frans de Waal, applying his research on chimpanzees to human behaviour, seeks to integrate and unify all the sciences according to evolutionary principles.[13] Wilson, trying to discern 'a deeper unity within the species', describes his theory of consilience as a metaphysical world view: 'Science offers the boldest metaphysics of the age ... there is a general explanation of the human condition proceeding from the deep history of genetic evolution.'[14]

Such beliefs are not theistic; they are not necessarily based on the existence of God or a spiritual entity. But they do follow a religious mindset that sees the world in terms of cosmic principles, ultimate purpose and design. Dawkins, who has been called the 'chief gladiator against religion', insists that anyone who believes in a creator, God, is 'scientifically illiterate'. 'Only the scientifically illiterate accept the why question where living creatures are concerned.'[15] He argues that the idea of higher purpose is an illusion and religion a dead issue. Yet Dawkins does finds ultimate purpose in human existence – the propagation of genes.

Wilson explicitly incorporates notions of purpose and design when he describes sociobiology as a science of systems design: 'If the theory of natural selection is really correct, an evolving species can be metaphorized as a communications engineer who tries to assemble as perfect a transmission device as the materials at hand permit.'[16] In *Consilience* he refers to people as 'adaptation executers'. Their adaptations are 'designed to maximize fitness, to exploit the local environment in the name of genetic self interest'.[17]

Though once a theist brought up to believe in God, Wilson now calls himself a deist, 'willing to buy the idea that some creative force determined the parameters of the universe when it began ... It would mean that human existence really is exalted and that immortality is a prospect.'[18] Though this force is not a God, Wilson's evolutionary epic purports to explain how the world works 'without surrendering the mystery of the Almighty and the need for communal liturgy'.[19]

Religious Rhetoric

Evolutionary psychologists have built their credibility on the success of molecular biologists in isolating disease genes. Convinced of the centrality of the genes, they believe that the mind will ultimately be reduced to material properties, that genetics has set the stage for understanding the still more complex systems of mind and behaviour.

The language used by geneticists to describe the genes is permeated with biblical imagery. Geneticists call the genome the 'Bible', the 'Book of Man' and the 'Holy Grail'. They convey an image of this molecular structure as more than a powerful biological entity: it is also a mystical force that defines the natural and moral order. And they project an idea of genetic essentialism, suggesting that by deciphering and decoding the molecular text they will be able to reconstruct the essence of human beings, unlock the key to human nature. As geneticist Walter Gilbert put it, understanding our genetic composition is the ultimate answer to the commandment 'know thyself'. Gilbert introduces his lectures on gene sequencing by pulling a compact disk from his pocket and announcing to his audience, 'This is you.' Former director of the Human Genome Project and Nobel Prize winner James Watson has proclaimed in public interviews that DNA is 'what makes us human', and that, 'in large measure, our fate is in our genes'.[20] And a student, writing in *The Pharos*, a medical journal, speculates, 'Given [its] essential roles in the origin, evolution and maintenance of life, it is tempting to wonder if this twisted sugar string of purine and pyrimidine base beads is, in fact, God.'[21]

Such images fuel popular narratives of genetic essentialism – a picture of the gene as the essence of the person, the locus of good and evil, the key to the 'secret of life'. At one level, the gene is a biological entity, the unit of heredity, a sequence of DNA that specifies the composition of a protein carrying the information that forms the tissues and cells. But it has also become a cultural icon, invested with social meaning and spiritual significance.[22]

The biblical references that geneticists use to describe DNA have buttressed the claims of evolutionary psychologists, who seek to move beyond molecular biology to reveal the 'hidden history of the prescriptive DNA stretched across countless generations'.[23] They too endow the gene with spiritual importance as a powerful and sacred object – an essential and immortal entity through which human life, history and fate can be explained and understood. They too elevate genes by treating them as a way to explore fundamental questions about human life, to define the essence of human existence and to imagine immortality.

Dawkins's extreme reductionism, in which DNA appears as immortal and the individual body as ultimately irrelevant, is in many ways a theological narrative: the things of this world (the body) do not matter, while the soul (DNA) lasts for ever. And Wilson says, 'you get a sense of immortality' as genes move on to future generations. Like the sacred texts of revealed religion, the 'evolutionary epic' explains our place in the world, our relationships, behaviour, morality and fate. It is indeed of truly epic proportions.

Missionary Fervour

Evolutionary psychologists are missionaries, advocating a set of principles that define the meaning of life and seeking to convert others to their beliefs. They are convinced they have insights into the human condition that must be accepted as truth. And their insights often come through revelations. Describing his conversion experience, Wilson notes that his biggest ideas happened 'within minutes . . . Those moments don't happen very often in a career, but they're climactic and exhilarating.'[24] He believes he is privy to 'new revelations of great moral importance', that from science 'new intimations of immortality can be drawn and a new mythos evolved'. Convinced that evolutionary explanations should prevail over all other beliefs, he seeks conversions.

Missionaries, inspired by their revelations, often place limited value on empirical evidence. Holistic narratives become more important than detailed logical structure, for theories follow from a kind of revealed truth. Evolutionary psychologists admit there is a paucity of examples for behavioural genetics. And they acknowledge the great difficulty in showing the empirical basis of epigenetic rules – the hereditary regularities in the development process – as applied to human behaviour. The theory of genetic fitness, writes Wilson, is supported by 'a scarcity of information' and 'the epigenetic rules that guide behavioral development are largely unexplored'. He admits that these shortcomings are conceptual, technical and deep, 'but they are ultimately solvable'. Trust, he says, is wisely placed in the 'natural consilience of the disciplines now addressing the connection between heredity and culture, even if support for it is accumulating slowly in bits and pieces'.[25] For it is 'better to steer by a lodestar than to drift across a meaningless sea'.[26]

Missionaries also tend to dismiss their critics. Evolutionary psychologists reject all postmodern thought, a category in which they include Afrocentrism, constructive social anthropology, eco-feminism, deep ecology, neo-Marxism and New Age holism. They label non-believers unenlightened, misguided, ignorant, unwilling to learn the truth, deluded, ideological or politically correct. They regard their critics as hostile forces, an image held over from the robustly belligerent response to sociobiology when Wilson first promulgated his ideas in the 1970s – a period less receptive to biological explanations of behaviour. These days, however, theories about the biological bases of human behaviour enjoy greater public and media support. But evolutionary psychologists are still frustrated by the reluctance of social scientists to adopt their models, and accuse them of 'tribal devotion to past masters and ideological commitments', of having a 'left wing political axe to grind'.[27] (See Hilary Rose's chapter.) Deluded and unenlightened beliefs about human behaviour, they believe, are more than a theoretical problem; they obstruct effective and moral social action.

Evolution as a Guide to Moral Behaviour and Policy Agendas

Evolutionary psychology is not only a new science, it is a vision of morality and social order, a guide to moral behaviour and policy agendas. By attributing human behaviour to the occult operations of the cell, evolutionary explanations lift behaviour out of the social context, denying the influence of human agency. And by defining behaviour as 'natural' – the consequence of evolutionary adaptations – these explanations convey a message about appropriate social policies. Evolutionary psychologists call for 'realism' based on the principle that behaviour is mediated by evolutionary forces.

Robert Wright argues, for example, that the idea of moral responsibility underlying the current legal system is outmoded and obsolete. Assumptions about moral responsibility are historically grounded in the premise that most individuals can choose freely how they will behave. But individuals cannot control what they do if they are driven to act by biological predispositions. Blame, then, becomes an unrealistic and intellectually groundless notion.[28] And, Wright says, arguments about intention, human agency, and free will are also meaningless when behaviour is reducible to evolutionary impulse. Policies must change accordingly: 'Tortured legal doctrines that defy ... our emerging comprehension of human nature ... are unlikely to withstand the test of time.'[29]

Evolutionary psychologists explain international violence in terms of evolutionary pressures among males. Warfare, they claim, can be understood as an adaptive strategy for acquiring the resources to mate and produce offspring that will carry on their genetic endowment. Evolutionary theories, they say, yield tools for identifying regions ripe for conflict before trouble actually breaks out.[30]

Evolutionary psychologists also apply their theories to the explanation of gender differences and to prescriptions about appropriate moral behaviour. Robin Baker believes that moral evaluations and realistic policies must take into account the differences between males and females which have evolved from the need to ensure that the fittest genes are carried to the next generation.[31] Women's natural abilities will lead them to prefer childcare to work outside the home. Richard Dawkins takes this idea further, to claim that women have a disproportionate stake in children because of their 'biological investment' of both time and cytoplasm (the egg is larger than the sperm). Differences follow from the 'abstract forces of evolution'.[32]

Such arguments, dealing with popular stereotypes, quickly reach the public through the mass media. In 1995 ABC aired a news special called 'Boys and Girls are Different' to announce the scientific evidence demonstrating the genetic differences between the sexes. Men have better spatial and directional abilities; women are better nurturers; men are better at maths; women have better verbal skills. These differences in intellectual and emotional skills, claimed the television host, had developed to assure evolutionary advantage. He concluded that the failure of women to achieve economic and professional parity with men was a consequence of these genetic differences rather than social and political forces. (See Anne Fausto-Sterling's chapter.)

Evolutionary explanations combine the credibility of science with the certainty of religion. They are especially convenient at a time when governments, faced with cost constraints, are seeking to dismantle the welfare state.[33] Why support job training, welfare or childcare programmes when those targeted are biologically incapable of benefiting from the effort? Theories about the evolutionary basis of status distinctions are a way to explain persistent poverty and social inequalities. Attaining status, so the argument goes, enables people to attract mates and to pass on their genes. Richard Herrnstein and Charles Murray developed the policy implications of this theory in *The*

Bell Curve, in which they argued that economic inequities are a ratification of 'genetic justice'.[34] Similarly, J. Philippe Rushton presents a theory of racial differences in brain and genital size, which he claims are based on evolutionary adaptations. Racial variation in skills, he says, is a consequence of evolutionary pressures.[35]

Those seeking to restrict immigration have found such theories useful. Some have claimed in policy debates that evolutionary pressures have led to the biological inferiority of some races and even nations. Cultural traits, they say, reflect genetic differences developed through evolutionary and adaptive changes.[36] For example, Peter Brimelow in *Alien Nation* writes that the process by which nations are created 'is not merely cultural but to a considerable extent biological'.[37] The policy message, received and promulgated in the mass media, is clear: 'Moral codes and policy prescriptions that don't acknowledge human nature are doomed to fail.'[38]

Nigel Nicholson, a professor at the London Business School and an evolutionary psychologist, has applied the evolutionary principle to organisational behaviour. He says in an interview in *Fortune* that he can show that certain companies are more successful than others because they have followed a model 'for which we were designed'.[39] And a *Business Week* writer attributes the ascension of Newt Gingrich to a position of political influence to his interest in the Dutch primatologist Frans de Waal, who studies power within a community of chimpanzees. Gingrich was an avid follower of this work on evolutionary behaviour and strategically applied its principles in the political arena.[40] But clearly there were deeper political forces which ultimately sealed Newt's fate.

The appeal of evolutionary psychology is, in part, politically driven. Evolutionary principles imply genetic destiny. They de-emphasise the influence of social circumstances, for there are natural limits constraining individuals. The moral? No possible social system, educational or nurturing plan can change the status quo. Evolution, defined as an eternal principle 'writ large', becomes a way to justify existing social categories and to deflect critical examination of the powers underlying social policy.

More than a scientific theory, evolutionary psychology is a quasi-religious narrative, providing a simple and compelling answer to complex and enduring questions concerning the cause of good and evil, the basis of moral responsibility and age-old questions about the nature of human nature. While represented as a scientific theory, evolutionary psychology is rooted in a religious impulse to explain the meaning of life.

Science and the Religious Impulse

Scientists have long argued that science and religion are separate and distinct spheres of life and that the appropriate relationship is one of mutual tolerance. Individuals, after all, are able to operate comfortably in both domains, even if their beliefs are philosophically in contradiction. Indeed, many scientists hold devout religious beliefs in their personal lives yet remain active in their laboratories. An often-cited example is Francis Collins, director of the Human Genome Project. He is an evangelical Christian who finds no conflict between religion and science; they simply operate in different spheres. He does find religious implications in his work which he feels gives him a kind of privileged entry into divine knowledge. When something new is revealed about the human genome, 'I experience a sense of awe at the realization that humanity now knows something only God knew before. It is a deeply moving sensation that helps me appreciate the spiritual side of life.'[41] Collins sees no conflict between scientific understanding of evolution and the idea of a creator God: 'Why couldn't God have used the mechanism of evolution to create?'

The 'no-conflict' theory is rejected by some historians. William Provine, historian of biology from Cornell, argues that the conflict between science and religion is fundamental and profound, and that the traditional truce between science and religion, based on the assumption that they deal with distinct domains, has been a convenient but unrealistic myth.[42]

Historian David Noble, in his book *The Religion of Technology*, takes issue with the very distinction between religion and science, demonstrating that science and technology have long been driven by 'spiritual yearnings for supernatural redemption'. Early science was a religious endeavour devoted to bringing man closer to the Divine Being. As 'handmaiden to revelation and prophecy', it is invested with spiritual significance.[43]

Today the religious impulse appears to be reviving. Having cut off the dialogue between science and the humanities through the 'science wars', many scientists are seeking to create a dialogue between science and religion. The fervent claims of evolutionary psychologists, their search for ultimate meaning through their theories, are not unique. God-talk has come into vogue among scientists who are using explicit religious metaphors to project the meaning and power of their work. A spate of recent books blur the traditional boundaries between science and religion when their authors claim, with nearly religious conviction and clearly religious language, that they have found the ultimate unifying principle which will reveal the most fundamental truths.

Science popularisations are full of all-encompassing explanations and spiritual claims. Science books have cosmic titles: *The Web of Life*, *The Physics of Immortality*, *The God Particle*, *Dreams of a Final Theory*, *The Sacred Depths of Nature*, *The Science of God*, *The Engine of Reason: The Seat of the Soul*, *The Astonishing Hypothesis: The Scientific Search for the Soul*, *Nature's Mind*, *Evolution*, *Morality and the Meaning of Life*, *The Biology of Moral Systems*.

Physicists were the first to describe their work in such cosmic terms. Leon Lederman, Nobel Prize-winning physicist, has named the sub-atomic entity which he believes determines everything the 'God particle'. Nobel Prize winner Steven Weinberg, in *Dreams of a Final Theory*, searches for the final principles that would explain all the laws of nature. Stephen Hawking, in *A Brief History of Time*, proclaims that scientists reveal 'the mind of God'. Physicist George Smoot has compared the big bang theory to 'the driving mechanism for the universe, and isn't that what God is?' Indeed, some cosmologists and physicists – and now evolutionary psychologists – sound like theologians seeking final answers to ultimate mysteries.

Meanwhile, organisations such as the John Templeton Foundation and the Center for Theology and the Natural Sciences (CTNS) have been formed to reconcile differences between science and religion and to foster 'mutual interaction'. In June 1998 the Templeton Foundation organised a conference called 'Science and the Spiritual Quest' to explore the common ground between science and religion.

The CTNS has over 500 members. According to spokesman Charles Townes, who shared the 1964 Nobel Prize for discovering the principles underlying the laser, the members sense a growing sympathy for religion within the scientific community: scientists have a 'growing tolerance for . . . a hierarchy of explanations in the world of the natural sciences'.[44] They believe it is a time for increased dialogue between religion and science. But many CTNS members also insist that contemporary science demonstrates there must be an ultimate purpose, a cosmic deity: 'Evolution occurs because all of nature is being grasped by the future that we call "God".'[45] A *Newsweek* interviewer describes their belief: 'The achievements of science offer support for spirituality and hints of the very nature of God.'[46]

In the mid-1990s, the ideas of evolutionary psychology spawned a quasi-religious organisation called The Epic of Evolution Society, attracting scientists, theologians and others concerned about spiritual values in the scientific age. Its mission is to activate awareness of the 'evolutionary narrative of the universe, life and humanity'.[47] In its website newsletter members describe evolutionary theory as 'a sacred story' with 'spiritual meanings', 'a tool for spiritual grounding', 'Darwin's great gift to theology', and 'a source of spiritual vitality in the world of consumer tech'. The editor claims that many members of the society have re-identified themselves as theists, 'bonded to the story of creation in spiritual ways'.[48]

Efforts to create a dialogue between religion and science, especially those sponsored by scientific organisations such as the American Association for the Advancement of Science (AAAS), in part reflect concerns about the influence of religious critics on the future of scientific research. Scientific creationists and other anti-evolution forces, seeking equal time for creation theory in the science curriculum of US public schools, have had significant influence on high-school classrooms and biology textbooks. Scientists also face the concern of religious groups that research in genetics and especially genetic engineering is desacralising the body, that scientists are 'tampering' with genes. Anti-abortionist groups have succeeded in blocking federal funding for foetal and embryo research.

Reasoned debate has not mitigated the perennial struggles between science and religion. The religious impulse among scientists – the God talk, the cosmic claims, the organisations for dialogue and reconciliation – may, to use their own favoured metaphor, be an adaptive strategy, a way to minimise the distance between science and religion. It is also, as evolutionary psychologists and their publishers have found out, a way to market books and ideas to a broader, non-scientific public in a society, notably the USA, where religion plays a powerful role.

In this context, it may be no coincidence that the depictions of genes in evolutionary narratives draw on powerful images of Christianity. Like the physicists engaged in God-talk, geneticists and evolutionary psychologists are borrowing the compelling concepts of one belief system to meet the needs of another, in an effort to attract converts – to convince the public and sceptics from other disciplines of the centrality and power of their ideas. But as scientists move from investigating how the world works to exploring questions of why – addressing age-old questions of what it means to be human, the nature of good and evil – they may only exacerbate the tensions between science and religion. Religious groups may seek a dialogue with science, but they are not about to step aside and leave it all to the quasi-religious narrative of evolutionary psychology.

Notes and References

1 E. O. Wilson, *Consilience: The Unity of Knowledge* (New York, Knopf, 1998), p. 4.

2 Ibid., p. 6.

3 *Random House Dictionary*, unabridged edition (New York, 1967).

4 AAAS Program of Dialogue between Science and Religion, brochure (1995).

5 Wilson, *Consilience*, pp. 168–73.

6 For a perceptive examination of unification theories, see V. B. Smocovitis, 'Unifying Biology', *Journal of the History of Biology*, 25 (Spring 1992).

7 The phrase is from Sarah Franklin, anthropologist, at Lancaster University.

8 See, for example, Thomas Bouchard et al. 'Sources of Human Psychological Differences', *Science*, 250 (1990), p. 223.

9 Wilson, *Consilience*, p. 129.

10 Robert Wright, *The Moral Animal: Why We Are the Way We Are. The New Science of Evolutionary Psychology* (New York, Pantheon, 1994).

11 Robin Baker, *Sperm Wars: The Science of Sex* (New York, Basic Books, 1996). Also see David Buss, *Evolution of Desire* (New York, Basic Books, 1994).

12 Richard Dawkins, *The Selfish Gene* (New York, Oxford University Press, 1976).

13 Frans de Waal, *Peacemaking Among Primates* (Cambridge, Harvard University Press, 1989).

14 Wilson, *Consilience*, p. 12.

15 Quoted in Greg Easterbrook, 'Science and God: A Warming Trend', *Science*, 277 (1997), pp. 892–3.

16 Wilson, *Sociobiology: The New Synthesis* (Cambridge, MA, Harvard University Press, 1975), p. 240.

17 Wilson, *Consilience*, p. 225.

18 Interview with E. O. Wilson in *Psychology Today* (1 September 1998), pp. 50ff.

19 Ibid.

20 L. Jaroff, 'The Gene Hunt: Scientists Launch a $3 Billion Project to Map the Chromosomes and Decipher the Complete Understanding for Making a Human Being', *Time* (20 March 1989), pp. 62–71.

21 Gregory Stephen Henderson, 'Is DNA God?', *The Pharos*, Journal of the Alpha Omega Alpha Honor Medical Society (Winter 1988), pp. 2–6.

22 Dorothy Nelkin and M. Susan Lindee, *The DNA Mystique: The Gene as Cultural Icon* (New York, W. H. Freeman, 1995).

23 Wilson, *Consilience*, p. 68.

24 Interview in *Psychology Today*.

25 Wilson, *Consilience*, p. 173.

26 Quoted in Easterbrook, 'Science and God'.

27 See, for example, Steven Pinker's comments in the *Houston Chronicle* (14 December 1997), p. 27.

28 Wright, *The Moral Animal*, p. 357.

29 Ibid., p. 203.

30 C. G. Mesquida and N. I. Weiner, quoted by Richard Saltus, 'Researchers Advance a Revolutionary Explanation of War', *San José Mercury News* (29 September 1998).

31 Baker, *Sperm Wars*.

32 Dawkins, *The Selfish Gene*, pp. 151–78.

33 See Dorothy Nelkin, 'Behavioral Genetics and Dismantling the Welfare State', in R. Carson and M. Rothstein (ed.), *Behavioral Genetics* (Baltimore, Johns Hopkins Press, 1999); Hilary Rose and Steven Rose, 'Moving Right out of Welfare and the Way Back', *Critical Social Policy*, 2 (1992), pp. 7–18.

34 Richard Herrnstein and Charles Murray, *The Bell Curve* (New York, The Free Press, 1994).

35 J. Philippe Rushton, *Race, Evolution and Behavior* (New Brunswick, Transaction Books, 1994).

36 Dorothy Nelkin and Mark Michaels, 'Biological Categories and Border Controls', *International Journal of Sociology and Social Policy*, 18 (1998), pp. 35–63.

37 Peter Brimelow, *Alien Nation* (New York, Random House, 1994).

38 Anthony Flint, 'Do We Still Think Like Stone-agers?', *Boston Globe* (21 August 1995), p. 25.

39 Geoffrey Colvin, 'Smart Managing', *Fortune* (16 August 1998), p. 213.

40 Elisabeth Leslie, 'Manager See, Manager Do', *Business Week* (3 April 1995), p. 90.

41 Quoted in Easterbrook, 'Science and God', p. 892.

42 William Provine, 'Scientists, Face it! Science and Religion Are Incompatible', *The Scientist* (5 September 1988), p. 10.

43 David Noble, *The Religion of Technology: The Divinity of Man and the Spirit of Invention* (New York, Knopf, 1997).

44 CTNS *Bulletin* (Winter 1998), p. 29.

45 Ibid., p. 5.

46 Sharon Begley, 'Science Finds God', *Newsweek* (20 July 1998).

47 Ibid.

48 Interview with Connie Barlow, editor of the society's *Newsletter* (21 November 1998).

3

EP, Phone Home

Charles Jencks

Ever since the cave dwellers of Lascaux and Chauvet posed the question 'How do we relate to the rest of the animal kingdom?' *Homo sapiens* has been divided on the answer. Are we driven by basic instincts, the Hollywood three – sex, power and money? Or are we ruled by our higher cultural faculties, the Platonic three – truth, beauty and goodness? Cro-Magnon people were exercised, as we are, by this perennial question, as can be seen in their beautiful cave paintings. These contrasted images of ferocious mammoths, bears and bison – nature red in tooth and claw – with paintings of little stick figures running about, noble, if Giacometti-like portraits, of shamans casting spells or groups dancing in order to propitiate the spirits of nature – make the rain fall or the Ice Age recede. There you had it, 30,000 years ago, the basic alternatives.

Their realistic art posed the metaphysical question in the starkest terms: is the world ruled by animal nature or human spirit? Or, as we'd say today, genes or cultural environment? As can be seen by studies of their art, comparing the number of animals versus the number of stick figures, they were evenly divided on the question. These are images of the human presence, often found at the entrance of caves or connected to sacred animals. Invariably they are produced either by stencilling or by imprinting directly in blood, so important was the question. If you decode these images, then you know that they wanted to know what we want to know – and for the last 50,000 years *Homo sapiens* has wanted to know: which is it? Nature or nurture? Now it appears that evolutionary psychology, EP, will give the answer.

That every age asks the basic question is beyond dispute. Why else the prevalence of sympathetic magic in prehistoric cultures? Why else the hybrid creatures that dominated Egyptian society for 3,000 years – the sphinx, the falcon-god Horus, the dog-god Anubis, and so on through many half-man/half-animal figures which populate temple walls? Why else have philosophers since Aristotle debated whether 'man the rational animal' is more animal or rational? Clearly we have this built-in genetic programme that demands we ask this question in every epoch and, since it is hard-wired into our brains by natural selection, we are bound to ask that Evolutionary Psychology tells us. EP, please phone home.

But let us adopt a little Platonic restraint and examine our *desire* for an answer before we give in to it, as we must in the end. Why are we so driven by such questions yet only partially satisfied by the answers we get? What evolutionary purpose does this serve? Why do we seek to *know* that our inner nature, or 'blood' as it used to be called, determines our action? Why, under Calvin or Knox, were people so happy to learn they were predetermined to be good or bad? Why, to bring the question up to date, are many neo-Thatcherites so happy to find they are led by their selfish genes, or conversely, why are so many anti-Thatcherites happy to find they are led by social and environmental forces – the zeitgeist red in tooth and claw?

These rhetorical questions revolve around our credulity, our need to find causal explanations of behaviour, as if people felt better knowing they are driven – by genes, nature or society. The desire for causation is so strong that it must be another example of hard-wiring – that is, a gene for wishing to be predestined.

Yet there is a conflict here. However much we are pushed towards believing in predeterminism by our DNA, today this must also be regarded with scepticism. Today doubt reigns everywhere. As the French philosopher Jean-François Lyotard has insisted, the postmodern age is a period when everyone is sceptical of grand theories, even those of science, and especially the one that explains this scepticism – postmodernism itself. So we are in a bit of a trap: nature predetermines us to want to know about causation and nurture, while the spirit of the age, postmodernism, predetermines us to be dubious of all metanarratives. What to do as these forces clash over us by night? There is a way out. As the Rabbi famously said to his son, 'Son, whenever faced with two extremes, always pick a third.' What could this be? Neither nature nor nurture? 'Neither Marx nor Jesus', as a recent French *philosophe* has it? Are we really to reject the only two alternatives the Nobel Laureate and geneticist Jacques Monod claimed were available: Chance or Necessity? Either Bad Luck or Bad Genes, as another book put the stark choice? Does that mean that it is neither sociobiology nor punctuated equilibrium which is in charge? What's left? EP, *please* phone home!

How did we get into this dualistic mess in the first place? These sort of questions fought it out in my soft-wiring and I was determined to find out, so I set out on a little voyage of discovery, reading books, going to conferences, interviewing experts and meeting Mr Darwin himself – or at least the man Tom Wolfe says is Darwin today. But more of that later. Besides the awful choice of dualism, the fault of Aristotelian logic, there is the great Western tradition from Plato to Descartes, the great 'either/or', and because it is really Cartesianism and the French who are to blame the most we might look to recent French philosophy to get us out of the quagmire. As everyone over sixteen knows, it is the Existentialists who made such a fuss with the answer to the dilemma: 'We are condemned to be free.' Freedom, obviously that's the third way, *not* the third way of Tony Blair nor that of the Fascists (who coined the phrase in the 1930s), but the third way of romantic novelists, Nietzscheans and complexity theorists. It is the last who tell us that freedom can emerge when self-organising systems are placed between competing orders.

We are, as they say, like any complex adaptive system, balanced on the 'edge of chaos', where creativity is maximised, where most alternatives can emerge. I won't trouble with the maths, nor the fractals, nor the examples of spontaneous emergence in systems pushed far from equilibrium, because there is already too much written on the subject.[1] But I will propose a logical diagram and typology. It is adapted from the familiar wave and particle duality of the atom and it can, in spite of its simplicity, shed light on this perennial conundrum. All the atomic elements, all 250 particles in the atomic zoo, from electrons to the Z particle, show the same complementarity. This produces a trade-off; the more that is known about the wave aspect of an electron the less is known about its existence as a particle, and vice versa. The same is true for their position and momentum, and some other complementary pairs. The complementarity or trade-off can be drawn as in figure 1.

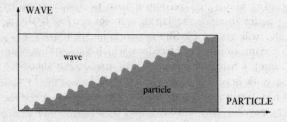

Figure 1 The Wave/Particle Duality

But, to remember the Rabbi's advice, we actually have *three* variables: nature, nurture and self-organisation. For convenience I will label them genes (G), culture (C) and free will (F). Let us look at some key human activities across the spectrum and at the extremes. A most basic predetermined action is the sneeze. Nearly all sneezes are involuntary and follow the usual pattern of sneaking up on you until you let fly with an 'A—CHUU'. Here nature and programmed nature – genes – are at work in the nose, mouth and larynx, so I would draw the relative influence with a big G-force, as in figure 2.

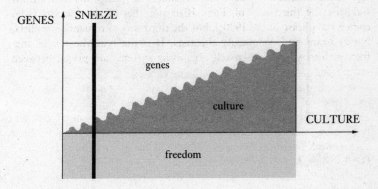

Figure 2 Gene/Culture/Freedom Triplet

However, it is said that when Germans sneeze they say '*gesundheit*', and so we must have a little place for culture (small c), and when my friend Louisa sneezes, she shows her individuality, or idiolect, by taking into her basic 'A—CHUU' a tiny, but recognisable suffix, 'tissue'. This represents two things: one half is an English, family articulation – more culture (small c), plus a little freedom actually to say 'tissue' rather than, say, 'A-SO', as for instance the Japanese might sneeze. If she is angry with me she will sneeze 'A-Shuuu', meaning 'get lost', a perfectly acceptable variant, showing her freedom, which I know how to interpret (showing mine). I hate to belabour a little sneeze, but there is a great principle at work here that is generally overlooked. (See Tim Ingold's chapter for a further working out of this argument.) Virtually all evolutionary psychologists attribute the 'tissue' in Louisa's sneeze to cultural conditioning, that is they turn it into another form of determinism, but the issue of the tissue is the sound which is found in the nose, I suppose. What? The point, as the previous sentence shows, is that the expressive plane always has a degree of individual freedom that transcends existing culture, because it represents something new.

Moving right along the bottom x-axis, we jump a few degrees of freedom to the next category, sex, the subject that novelists say rules

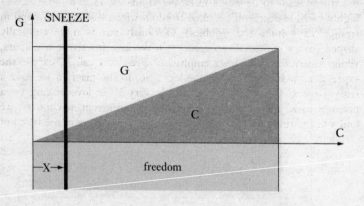

Figure 3 How Free Is a Sneeze?

people's lives: one of the three main determinants of 'sex, power and money', the three eternal subjects. As we are too well aware, Clinton got the first, sex (S), because he had the second, power (P), and as a consequence he lost the third, money (M), all of which can be mathematicised as $P - S = -M$. Or, moving the integers across: $+ S$ and $+ M = - P$. QED, he's (nearly) impeached. Excuse the rather mechanical determinism, but sex *is* important in the spectrum of behaviour for showing a certain fluctuation between rigid repetition and unique peak moments. And these are not idle comments, or some cheap Clinton joke found on the Internet, but reflect serious considerations that go to the heart of the 30,000 year debate. Consider *Time* magazine's line of defence, 'The "It Could be Me" Factor':

> Many women, mulling Clinton's sins, don't ask how they would have acted in his situation but how they would have acted if burdened with male genes – and, perhaps, with a sense of entitlement initiated by years of alpha maledom . . . genes for large appetite . . . Genes and environment, so far as science can tell, are all there is.[2]

All there is! What kind of science is this? You see how the readers of
Time have been stunted in their outlook, trapped between the two
determinisms, as if that were all there is. Like evolutionary psychologists, they have yet to break through the adolescent/existentialist
barrier. Figure 4 brings out what is involved.

You see that there is the usual trade-off of nature and nurture, except
that there seem to be many more determinants on both levels *and* a
highly variable degree of personal freedom which distorts the graph in
strange ways (note the wobbles). Obviously sex is more culturally
shaped than a sneeze. Tantric adepts make love for six or ten hours,
while Americans sometimes emphasise speed and call it, excuse the
expression, 'a sneeze between the legs'. So, in the diagram we have a
fairly large C-factor. Since individuals vary their love-making on a
personal basis, and as they go along with it and invent new moves, we
find a fluctuating blip of F-factor: freedom. Here we are in the realm of
creativity, self-organising systems, art, individual expression, and what
Steven Rose calls 'lifelines', that is, the personal history constructed
within the constraints of biology, society and all the rest of it.

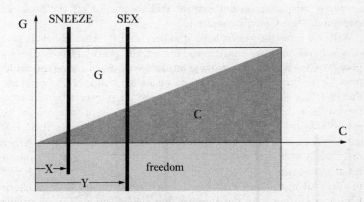

Figure 4 Sex Is Freer Than a Sneeze

As we move along through more essential activities such as eating,
playing and dying, it becomes clear that there are lots of blips and
bumps along the graph. The trade-off is a lot more complex than I've
allowed, with feedback between the three areas and probably much else
besides. But I want to keep things simple with the graph because it does

illuminate a major point, which I'll call the Sartre Quotient, after the existentialist Jean-Paul Sartre. Recall that famous incident with Sartre, Simone de Beauvoir and a third party (not Camus) sitting around in the Café Deux Magots in Paris. Their strenuous arguments about freedom suddenly led in an unexpected direction, to the question of whether they could order another coffee by calling the waiter '*garçon*'. To treat him as a boy, a lackey and paid servant, was to take his freedom away, reduce him to the level of a conditioned object. Thus, or so a version of the story goes, they could not order coffee. This illustrates that the freedom quotient is highly variable in certain activities and dependent on interpretation, on the meaning invested in certain words: also it is proportional to the degree that people care to exercise it.

Nowhere is this Sartre Quotient higher than in a work of art, a novel, or a poem. It is also quite high in a work of architecture, although less than in painting or writing because of the great social, economic and material constraints. What happens to our diagram when we look to the extreme of art – if I may call it an extreme, because freedom is at a maximum here? Notoriously it is the area where evolutionary psychologists, such as Steven Pinker, become vague, where evolutionary archaeologists, such as Steven Mithen, invoke Stephen Jay Gould's 'spandrels', the architectural element said to get one off the hook of adaptation. (See Gould's chapter.).

Well, to pose the really hard question to EP, what about James Joyce's *Ulysses* or, less idiosyncratic but equally poetic, Dante's *Divine Comedy*? Clearly, like all art, these works are as much concerned with

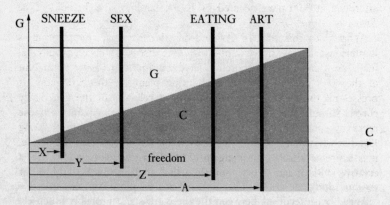

Figure 5 Art Is Freer Than Sex: The Expressive Plane

expression as with content. The sound and associations of poetic language are as important as its explicit message. No adaptation there. But we want to know how to *measure* the degree of creative freedom in a work of art as against the degree of adaptational function. Does that sound absurd? Well, many aestheticians and not a few scientists have attempted to quantify beauty and the information content of syntactic and semantic elements and draw some conclusions. The basic idea is that the more complex a work is *and* the more it manipulates basic built-in rules of perception, then the more successful and the better it is.

The neurophysiologist Semir Zeki is an expert on the way the brain processes different syntactic elements. He has proposed some constructional rules that may underlie good painting. Not surprisingly, he finds that the better paintings seem to resemble the Constructivists' work of the 1920s, a time when abstract syntactic forms were partly the subject of art. Clearly, rules of composition are important. Formal coherence intensifies the qualities of a painting, and some syntactic elements are hard-wired into the brain, to a degree, as they are in language. But, just as obviously, some *semantic* elements are also built in, and, in conversation, I mentioned to Zeki the amusing case of the contemporary Russian-American painters Komar and Melamid. These two mad-satiric inventors set out to paint the greatest, most popular painting ever painted and, in a straightforward way, researched all the previous most popular paintings ever painted and then, quite naturally, did a composite of all of them.

The result, if I remember rightly, resembled a combination of a Constable landscape, a Currier and Ives print of the American landscape, a seascape, mountainscape and cityscape, all seen on a sunny afternoon at 3.00 p.m. peopled by happy families straight out of *The Sound of Music*.

What does this tell us? Aside from the fact that the painting is lugubriously funny, it does appeal to basic instincts, those adaptational just-so stories that are supposed to have conditioned us as we swung out of the trees and landed on the hunting plains of the savannahs. It presses all the right functional buttons, it releases all the genetically correct stimuli. We respond to it as we might to Sunday painting, those racks of sentimental images suspended on the park railings along London's Bayswater Road. But, as even EP might admit on the phone, it is supreme kitsch. An amalgam of all the right releasers, but not a creative work of art: a cliché soufflé, which works as high pastiche but fails as high art.

What is lacking, among many things, is depth, creative complexity, the interrelationship of new meanings. How do we measure these things, even approximately? Tito Arecchi and others involved with complexity theory have worked out very crude analogues for measuring such depth. Arecchi gives as an example the sequence of vowels from the first canto of Dante's *Divine Comedy*. He contrasts this with (I) a random sequence *aoeeaaoae*, and then (II) a regular sequence *aeoaeoaeo*. The *Divine Comedy* sequence is (III) *eeoeaoaaoaeaeaoaeaaaeaaa*. His comment:

> Let us agree to define as complexity the cost of the computer programme (length of the instructions multiplied by processing time) which enables us to realise one of the three sequences. In the case of sequence I the programme is simple: 'write at random a, or e, or o'. For II the instruction is very short: in III there exists no programme which would be shorter than the Divine Comedy itself. Sequence III we may call complex, the other two simple.[3]

This is hardly the last word on measuring complexity, but it shows that what is at stake lies within the work of art and is part of its quality. Discussions of art which never broach such questions of organisational depth and quality never get very far with the vexed relations between culture and evolution.

Meeting Mr Darwin

My chance to explore these questions came when I was asked to speak at an international conference on postmodernism in Boston, in late November 1998. The pretext was an annual meeting of assorted experts perched momentarily, like migratory birds, en route between one country and the next – England, Australia, the States – wherever the intellectual ecology was most propitious for the chosen topic. These gatherings are called 'The Boston, Melbourne, Oxford Conversazioni on Culture and Society', and I thought nothing could be better for understanding the big issues than an open conversation, a discussion without *parti pris* (even if it had a slightly nineteenth-century ring with its ending flourish *zioni*). I was to represent, and defend, postmodernism in debate with the other speakers. It was clear where at least one of them, the most famous, stood – Professor Edward O. Wilson from Harvard – because he had just given his view in his 1998 biopotboiler *Consilience*: 'Postmodernism is the ultimate polar antithesis of the Enlightenment. The difference between the two extremes can be expressed roughly as follows: Enlightenment thinkers believe we can know everything, and radical postmodernists believe we can know nothing.'[4]

Naturally I wanted to have a gentle *conversazione* with him on this point, but really, deep down, I wanted to hear his views on human nature. He has been one of the most influential thinkers on the subject ever since he wrote his magisterial polemic *Sociobiology: The New Synthesis* in 1975. Since then he has been Mr Sociobiology, the Godfather evolutionary psychologist as it were, moving along the modernist trajectory in one continuous argument, becoming ever more developed and sharpened, from *On Human Nature*, 1978, to *Success and Dominance in Ecosystems: The Case of the Social Insects*, 1990, to *In Search of Nature*, 1996, right up to his latest, *Consilience*. His important work on one of the most social species of all, ants, has always informed the argument. But, so far as I was concerned, he was at the same time a closet postmodernist, because he also gives very passionate sermons on biodiversity. The southern Baptist in him gives his message bite, direction, persuasion. He wants to convert the world to his love of the bios – 'biophilia' – in order to save it from destruction, and if his passion won't win the argument then, he hopes, the evidence and logic will. Thus he is an attractive, complex sharpshooter with at least three bows to his quiver – ants, sociobiology, biodiversity – and a goal to convert the world to *Consilience* as expressed in its subtitle, *The Unity of Knowledge*, that Enlightenment ideal. His Harvard power, his influential books, his Pulitzer prizes, his Big View, led Tom Wolfe to launch him, on the cover of his latest tract, with the simple rocket: 'There's a new Darwin. His name is Edward O. Wilson.' He grins out from this cover impishly, a veritable Huckleberry Finn crossed with an avuncular Harvard professor. 'How nice', I grinned back, inwardly responding to several releasers, 'to meet Mr Darwin, even if he takes a dim view of postmodernism.'

The *conversazioni* progressed through its several stages, moving from 'Origins' to 'Consequences' and culminating, if that is the right word for rejecting the postmodernist claims, with 'Reconsiderations'. Wilson was the final speaker, his talk fittingly 'The Unity of Knowledge'. He ambled up to the lectern and took no time launching into his mission statement: 'There are only two ways to account for the human condition, that of the sciences . . . and all the other ways.'[5] The audience, largely non-scientists with a sprinkling of postmodernists, all of whom had just been rendered obsolete, nevertheless warmed to this thought because it was delivered with such benevolent good cheer. The ultimate unity of knowledge, presumed by scientists, is now moving forward to some far-off goal of consilience, a word coined in the 1840s and thought to be preferable to more common terms, such as integration, which carry too much baggage. We can see, just over the horizon, this goal of harmonious unification because of four emergent disciplines: cognitive neuroscience, behavioural genetics, evolutionary biology and sociobiology.

For a second a sceptical thought flitted across my neurons: 'What about the basic contradiction that exercises physicists – that between quantum and relativity theories?' 'Silly doubt,' another neuronal cluster told me, 'this small hurdle will be jumped in the next fifty years with a new synthesis. As supersymmetries are found, as superstrings become one type, as TOEs (Theories of Everything) another and GUTs (Grand Unified Theories) reconcile the lot, peace will reign for ever in the land of Consilience.'

Wilson then flashed on the screen one of those familiar slides of exponential growth. Not the negative ones of population explosion, mass extinctions and so on, but a positive harmonising one – 'genomic research'. The graph ascended at the far right, straight heavenward, and that led him into the Big Subject – Human Nature. 'Human Nature', he clarified, 'is not genes that prescribe it. Rather, Human Nature is based on the epigenetic rules that predispose us to act in certain universal ways. Some of these rules, millions of years old, tie us to the primates and chimps: these include our body language, our fears and phobias, our facial expressions such as smiling, which are recognised by all cultures with an accuracy of over 80 per cent. For instance, we have an innate aversion to snakes, as do monkeys and chimps, which give warning cries when they encounter poisonous ones. Children learn a fear of snakes *more easily* than they remain indifferent. A bias, or predisposition, or propensity, that is an inherited regularity, is an epigenetic rule – an algorithm of growth . . .'

Another shower of unconsilient photons shot through my grey matter. Another doubt. My eighteen-year-old daughter had just returned home having smuggled through customs a small pet snake. While I reacted with the correct humanoid bias – fear, disgust, horror – she actually draped this slimy, yuk-infested vermin around her neck like a string of pearls. Where was her natural reaction? Maybe, I thought, sliding deeper into my seat, she isn't human, or even a primate. 'Another epigenetic rule', he continued, ' is the incest taboo which is shared with monkeys and apes. We all have it in order to avoid inbreeding, genetic mistakes; it's millions of years old.' Suddenly, I felt better, relieved at the news, I share the aversion with my daughter; we rejoined the human race.

Professor Wilson enumerated several more examples of epigenetic rules at work as he advanced, stage by stage, towards the contentious area of art and culture. Human Nature tends to have certain colour biases built into it, certain 'natural categories' for cutting up the continuous spectrum of light waves into, say, black and white. With some twenty cultures studied, Wilson told us, a common cluster of colour areas emerged – blue, green, yellow and red – in spite of the fact that these clusters do not exist in nature. Remember, light spectra are continuously varying, they do not come in little tubes of paint. Even more astounding, it is now known, we *learn* and *combine* our colour categories in a certain regular fashion, and this must entail a deep epigenetic rule.

Moreover, and here we reached a kind of climactic insight not far from my own prejudices, there are kinds of syntactic rules that lead 'bioaestheticians' to classify preferences for certain degrees of complexity. The question is: how simple and repetitive, or complex and varied, should an abstract design be? Researchers found there is a 'peak of brain response when the redundancy – repetitiveness – in the designs was about 20 per cent'. Wilson showed ancient petroglyphs which had this degree of redundancy and built up his case with photos of grillwork, Japanese calligraphy, Punjabi scripts, flag designs and – yes – the best Mondrians. No question, maximal brain arousal peaks around this figure, 'the 20 per cent redundancy effect'.

In order to bang in the nail, three labyrinths were shown: a very simple order of repeated horizontal lines with a left-hand wall (little brain arousal); a deconstructionist one full of staccato stops and starts (too much brain arousal); and a nicely complex one with strong border and internal variety. Which was the best? The middle one of course, just the right amount of brain arousal, even I, a postmodernist with a snake-loving daughter, could see *that*. 'Twenty per cent redundancy! Of course, how had we missed it in twenty-five centuries of architectural aesthetics, the happy norm was staring us right in the eye-brain-arousal organ!'

But no, that damn photon-flicker of doubt zapped through my cranium, try as I would to suppress it: 'What about the Modern Masters? Did they have the right degree of redundancy? Of Simplicity divided by Complexity? What about "difficult art" where you had to work at it?' 'No, that's not right either,' my right hemisphere answered. 'There must be some limit on the degree of labyrinthine complication; Wilson must be right. Too bad James Joyce, too bad T. S. Eliot, your *Wasteland* is just too ... complex, not enough redundancy.' Thirty more masterpieces of the Modern canon flicked through my mind, sending it into further cognitive dissonance, this way and that, from Stravinsky to Jackson Pollock. But I suppressed such agitating doubts as Wilson turned to the subject that interests everybody, not just bioaestheticians. Measuring human beauty.

Apparently, we have known for more than a century that we prefer composite faces blended together; we consider them more attractive than individual faces viewed separately. You and I may not have known this strange truth, but researchers into beauty have known it. How? From morphing photographs together, by producing composite images, which always turn out to score the highest. The inference from this, we might think, is that we prefer the average of all the faces, and this would seem to have some evolutionary significance – adaptation favours the middle way, 20 per cent redundancy or conformity wins. The morphed face might serve any of these adaptive functions? 'But no,' Wilson grinned with that wry twinkle as if to say, 'you know evolutionary psychology is not as simple as *that* . . . Oh no, the ideal woman is *not* the exact average.'

Wilson then again took us through the slide test, with three morphed beautiful faces side by side. My neurons went into overdrive as I desperately tried to fathom which was the Ideal Beautiful Woman (IBW)? Middle one? Twenty per cent redundancy? No use, I couldn't find her, I could hardly even see the difference between the three. Wilson continued, barely suppressing pride at the subtle levels of counter-intuitive measurement that EP had reached: 'The real surprise of the tests is that certain characteristics of the face found to be attractive give considerably more beauty when they are exaggerated . . . high cheek bones, thin jaw, large eyes and a slightly shorter rather than longer distance between mouth and chin and between nose and chin . . . Few women approach this standard of facial beauty, which should be the case were it just the normal example of natural selection . . . Instead we have a *supernormal* exaggeration of those features which are signs of youth, virginity and the prospects of a long reproductive period . . . Ask', he flashed a triumphant grin, 'the American president!' And, with another wink at the audience, 'Look at the *second* wives of most academics!'

I slid down again in my seat and sneaked a look at all the academics around me: 87 per cent, average age fifty-two I guessed, maybe 32 per cent of them into second wives, what were *they* thinking? Did these second wives have higher than average cheekbones, thinner jaws and better signs of reproductive potential (poor unsuspecting academics, faced with those supernormal features, they didn't stand a chance).

As a wishful Alpha male, I had fathomed some attributes of reproductive potential, at least the more obvious ones that stare out at you from every magazine counter. My researches had even gone so far as to discover that, in a recent poll, *Playboy* had discovered *the* Ideal Beautiful Woman of the twentieth century. No marks for guessing it was Marilyn Monroe, who vastly outstripped all the others, even the second-placed Jayne Mansfield.[6] That poll showed what we now know: beauty relates to sex, which relates to reproductive potential, and as the evolutionary psychologists have conclusively shown, all this leads inexorably to what men like when they are little babies: big busts and lots of blonde hair. Such are the fruits of 30,000 years of progressive research. Tough question: Why do gentle-baby-men prefer blondes with big curves top and bottom? EP breakthrough: Because all these features send out childlike signals, of women like Marilyn, who are easy to cuddle and dominate – an old male fantasy of being young and reproductive again.

'But no! Again and a thousand times no!' My neurocircuitry just would not reach consilience on this score. 'What about all those Mother Earth goddesses who dominated European life from 9,000 to 2,000 BC? Wide hips, sagging breasts, massive thighs – not the hourglass but the pear figure. What about the Wilendorf Venus, the major icon of female beauty from 28,000 to 9,000 BC? What about the painter Boucher, and his roly-poly bottoms, or Rubens and his acres of sexy hanging flesh – or the opposite, the IBW of the Sixties, Twiggy? Or all those emaciated, I-am-on-coke-and-about-to-die models in *Vogue* (not much reproductive potential there)?' As I was mentally zigzagging through 35,000 years of female pulchritude, doing a factor analysis of the essential features of the IBW – and seeing them jump around like silly putty before my mind's eye – I was suddenly snapped out of disharmonious reverie as Wilson changed gear, and slides.

He showed an idyllic picture of a rolling landscape with meandering pools of water interspersed by shrubbery and a sprinkling of trees – all this nestled into a background of well-clipped lawns. Now he was in his element, real biophilia here, you could almost hear the birds sing, the worms turn. Why, the slide seemed to ask, do we all naturally respond to such an Elysium? Because, as evolutionary psychology has shown, there are certain universal features of a good place and these are searched out by Human Nature. What are its characteristics? It is a place on-high, perched over the landscape which one can survey like an eagle: good defence strategy here, naturally selected by natural selection and built firmly into our epigenetic rules. Second, the picture revealed our preference for flat open fields; again no mystery here because we spent millennia roaming the prairies after we swung out of the trees – every school child has heard that one. The third feature that constituted this image as Best Place was the proximity of water. Why? You guessed it, time for a drink, go have a bath, rather universal drives there. You don't have to be the average American – one bathroom per bedroom – to know that the Best Place (BP) is near water. Fourth and fifth qualities that we look for when we choose an ideal habitat are a retreat from others, a hideaway from strangers and mothers-in-law, and a place to forage, putter around for grubs.

Wilson showed further slides of this landscape and it turned out to be the John Deere Headquarters in the countryside, a place where this multinational corporation surveys the world market from on high. Of course, it was obvious all along; how silly of us not to have guessed before evolutionary psychology uncovered the truth: Human Nature is a Suburban Corporate Executive foraging around his back garden. Four million years of human evolution had come to this, no small thing, and evolutionary psychology has cracked the code, no minor triumph – though, as Wilson observed with a wry upper turn to the edges of his mouth, 'Real-estate dealers have understood these five points for a long time.'

Why was he smiling at this, why was he bubbling over with infectious enthusiasm, why did it make me feel so good? Even though I didn't agree with every word he said about postmodernism, his electric beam melted down all resistance until, again, suddenly – horror, a dark flash of dissonant doubt: 'Could this smile be that of the hostess or the real-estate dealer? Is it a supernormal stimulus I cannot resist?' He had written that smiling was fairly similar across cultures and that 'It invariably attracts an abundance of affection from attending adults.'[7] Then I remembered that a recent breakthrough in evolutionary psychology had carefully calibrated the difference between a true and false smile, the politician's and child's grin. So, concentrating hard on Wilson's upper ligature, I tried to decode the subtle differences in muscle tone around the eyes – where the truth of smiles is marked. Alas, from where I was sitting, all smiles looked alike.

Wilson continued with pictures close to his heart: views of the rainforest, the tangled jungle of biodiversity that appealed to him in fifty years of studying ants, and the kind of riotous variety that had also captivated Charles Darwin one hundred and fifty years before. There is something undeniably enticing about photographs of these rich ecologies, even with their redundancy quota below that 20 per cent optimum for brain arousal. Darwin was particularly fascinated by an etching of a Brazilian rainforest. This showed a lush undergrowth encompassing a plenitude of species, the kind of luxuriant mystery which makes the jungle not so much red in tooth and claw as like a well-stocked department store brimming over with Christmas gifts. 'There is an innate resonance of this with us,' Wilson observed, reminding us that we are Platonic tuning forks. 'The sublimity of this picture may relate to our attraction to explore the unknown.'[8] Yes, I could agree with that. The search for knowledge is almost as powerful as the sexual drive and I could see that the voluptuous, tangled jungle was somehow deeply attractive. Perhaps it released my exploration and curiosity hormones, or it was resonating with my universal desire to unify all the tangled strands of knowledge and tie them into one Enlightenment Theory of Everything?

But then, as he pressed the next button, I saw where this argument was leading him: to the neo-realist rendering of Alexis Rockman: paintings of a Guyana forest dripping with haloes of mist, shrouded with the kind of atmospherics that did lead the mind to explore, that did draw the imagination forward. But to what? Images slightly reminiscent of Max Parish's romantic musings? No. Or the films of Steven Spielberg's Dreamworks? Or Disney Imagineers? A bit more, but not quite. No, we *had* been there before, it was the great American world view of Norman Rockwell – applied to nature. That's where the argument was taking us, ineluctably. Just as the features of the Best Place pointed to a suburban corporate headquarters, just as the Ideal Beautiful Woman pointed towards a high-cheekboned Marilyn Monroe, the deep drive to explore the fecundity of nature was leading us to the art of the Bayswater Road. There is no escape from this conclusion, Wilson has the courage to face it several times in *Consilience*:

> Gene-culture coevolution is, I believe, the underlying process by which the brain evolved and the arts originated . . . This much can be said with confidence, however: The growing evidence of an overall structured and powerful human nature, channeling development of the mind, favours a more traditionalist view of the arts. The arts are not solely shaped by errant genius out of historical circumstances and idiosyncratic personal experience. The roots of their inspiration date back in deep history to the genetic origins of the human brain, and are permanent.[9]

Permanent! Well, Disney will be glad to hear this. It means it is Norman Rockwell for ever, and kitsch all the way down. This line of argument warmed some of the academics gathered at the *conversazioni*, particularly those on the platform leading the concluding session.

Together, with one voice, these seven wise men proceeded to flay postmodernists with versions of this message, in case anyone didn't get it. The conversation proved it: when it comes to the battle of *kulturkampf* the victory always goes to Lower-Mid Cult.

'But no!' the old wetware rang out somewhere deep within my reptilian brain, 'Hang on there, it just doesn't sound right. Granted it resembles heaven, but why should evolution stop and get fixed for ever in suburban America?' Surely, as the history of art shows, culture and taste have historical and contingent components. Surely, as women's cheekbones and vital statistics show, there are chaotic attractors which set taste in one place, and then another, depending on a lot of things: the economy, food, fashion, ideals of aesthetics, Dr Spock's theories of bringing up baby. Each age interacts with the previous one; all periods hunt for new beauty; all art at least partly focuses on the expressive plane, and there is something of profound importance here. There will always be a key part of the expressive language that is emergent, new, novel – and sexy for being so; and it will never be reducible to hard or soft-wiring within *Homo sapiens*. Every good artist has known this truth: that creativity concerns moves made within the language of art, independent of the neurocircuitry that responds to it.

I turned to the man sitting next to me, and whispered, 'Before the evening is out, I'll bet that Wilson justifies the moral direction which comes from his research. I bet he goes straight from an "is" to an "ought".' And as I left the dinner that night, Wilson came to the microphone and did precisely that: committed the logical error that Modernists, since David Hume, have been warned not to commit. The reason for this academic stricture became particularly apparent in the late nineteenth century, with Social Darwinists and John D. Rockefeller, who argued that, since nature shows the survival of the fittest coming out of competition, then society should make permanent the winners and losers. It is only natural to follow natural selection. In spite of such arguments continuously being shown to be logically false and morally suspect, they are, I believe, being continuously made and especially by those trained to avoid them, academics. The reason is not hard to find. We easily slip from what is likely, what is programmed by nature into us, what is an epigenetic rule, into believing it is something that cannot, in the long run, be resisted. This is, ultimately, the solecism that *Consilience* achieves, despite its author's disclaimers. A gene predisposing Clinton to certain behaviour will always be used as partial defence of that behaviour; a famous cleric in Scotland even declared that in an age in which we know that there is an epigenetic rule for infidelity, it could no longer be censured as it was in the biblical age. We slide all the time from an 'is' to an 'ought', no matter how many road signs we put up condemning the logical slip. And by the same token, of course, this is no reason to excuse the behaviour; it is merely to point out its common occurrence. Nowhere is it so prevalent as with evolutionary psychologists, who often study human behaviour and then, like Wilson, tell us to get in step with it; or, as he does, generalise about what most art has been and deduce therefore what it should be.

The long 30,000 year debate about who we are and what we can be is in no danger of ending. Our animal and vegetable natures do have propensities which guide us somewhat, just as they do an amoeba. But genes maketh not the man any more than clothes. Even EP-enthusiasts, such as Richard Dawkins and Steven Pinker, when they do not like the message, tell their genes to jump in the lake. At these rare moments, apparently, EP phoned in and said, 'You're free, but not completely,' and the message stayed on their answerphones (or at least remained in their books, inconsistent though it may be). If we are to take an unfolding universe seriously, we must acknowledge the deep truth of emergence and limited creativity. Novelty acknowledged as radically present means that we can never reduce behaviour to nature, nurture or their combination. With the arts, because of their concentration on the expressive plane, this limited creativity opens out exponentially. It leads to a final graph the artists of Lascaux would have understood.

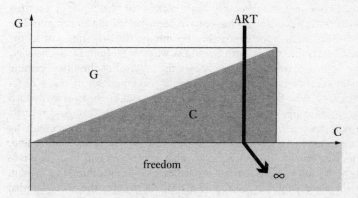

Figure 6 Rule of Increasing Freedom

Notes and References

1 See, for instance, Mitchell Waldrop, *Complexity: The Emerging Science at the Edge of Order and Chaos* (New York, Simon & Schuster, 1992); Roger Lewin, *Complexity: Life at the Edge of Chaos* (New York, Macmillan, 1992); Charles Jencks, *The Architecture of the Jumping Universe: How Complexity Science is Changing Architecture and Culture* (London, Academy, 1995 and 1997).

2 *Time* magazine (28 September 1998), pp. 78–9.

3 From *Liber 1*, *Times Literary Supplement* (6 October 1989), trans. R. G. Harrison.

4 E. O. Wilson, *Consilience: The Unity of Knowledge* (London, Little Brown, 1998), p. 40.

5 This and the other quotes are from my notes checked against *Consilience*. The subject areas I quote can be found on the following pages: epigenetic rules, pp. 150–3; aversion to snakes, p. 79; incest taboo, pp. 173–80, 194; colour bias and categories, pp. 159–63; monitoring brain arousal for 20 per cent redundancy effect, pp. 221–2; 229–30; the beauty of a woman's face, pp. 230–2.

6 The *Playboy* poll on sexy women and its expansion by *Nature* are discussed in Ros Miles, 'Sexy Women? Your Eyes Deceive You', *Sunday Times* (29 November 1998), p. 7.

7 Wilson, *Consilience*, p. 153.

8 Exploratory instinct, ibid., pp. 232–3.

9 Ibid., p. 218.

4

Anti-Dawkins

Gabriel Dover

The Case for Polemic

For readers of the Marxist classics, my title will come as no surprise. It is derived directly from Friedrich Engels's mischievous book *Anti-Dühring*, in which Herr Eugen Dühring's self-proclaimed 'revolution in philosophy, political economy and society' was systematically exposed for what it was. From the pieces of this exposure Engels managed to do two things: one intended, the other unintended. The first was a constructive account of an alternative analysis, and the second was the immortalisation of Herr Dühring, who would otherwise have gone the way of many another nineteenth-century European systems builder.

There are many similarities between Dawkins and Dühring and, like Engels, I intend to use an anti-Dawkins polemic as a device to paint a richer picture of some of the late twentieth-century excitements of the 'new genetics' and its evolutionary implications. These developments have no room for Dawkins's vulgar misappropriation of the theory of natural selection, as embodied in the selfish-gene delusion. There is a story here to be told in itself, how the 'meme' of the 'selfish gene' replicated and infected the studies of human psychology, animal behaviour, philosophy, sociology and medicine. But that is for the historians of the sciences to fathom one day. I'm prepared to say, in passing, that the initial impulse to accept the selfish-gene abstraction was based on a dangerous cocktail: mixing the seductive and relentless advocacy of Dawkins himself with an ignorance of the fundamental workings of the genetics of natural selection by non-biologists and science writers. It will be to the lasting shame of the seducers and seduced that an intellectual idleness pervades much that they seek to explain when confronted by issues of massive complexity. Consider how, for example, a single fertilised egg unfolds into a multicellular

47

organism, mirroring a three and a half thousand million year unfolding during evolution. The essential wrongfootedness of the selfish gene as a serious concept in evolutionary genetics, and the overwhelming ignorance of this failing in so many peripheral disciplines, are now of major concern.

Genes are so battered, misunderstood and abused that I make no apologies for starting from the beginning with the genetic material. Genes are not self-replicating entities; they are not eternal; they are not units of selection; they are not units of function; and they are not units of instruction. They are modular in construction and history; invariably redundant; each involved in a multitude of functions; and misbehave in a bizarre range of ways. They co-evolve intimately and interactively with each other through their protein and RNA products. They have no meaning outside their interactions, with regard to any adaptive feature of an individual: there are no one-to-one links between genes and complex traits. Genes are the units of inheritance but not the units of evolution: I shall argue that there are no 'units' of evolution as such because all units are constantly changing. They are intimately involved with the evolution of biological functions, but evolution is not about the natural selection of 'selfish' genes.

One Step Forward, Two Steps Back

Dawkins claims that he was appalled, back in the 1970s, by the prevalent sloppiness of animal behaviourists to explain natural selection on the basis that what was selected was 'for the good of the species', or 'for the good of the group'. Dawkins's discomfort over the supposed misappreciation of the mechanism of natural selection in his own discipline (zoology) is not sufficient justification, however, for substituting his own mistaken notion that natural selection is 'for the good of the gene'. The only attempt to accommodate the gene as the unit of selection concerns the phenomenon of altruism, as if this behavioural trait, considered by sociobiologists to be the last remaining unsolved problem in evolution, is living proof of the gene as the unit of selection. But, in the wider context of the living world of viruses, microbes, plants, fungi and the majority of animals, this phenomenon is marginal in the extreme and can, in any event, be explained by alternative models of evolutionary group selection.[1] Indeed, Dawkins is after a much larger prize than providing lyrical support to the theoretical underpinning of kin selection and altruism as originally developed by William Hamilton.[2]

He is pursuing a global explanation for the evolution of complex adaptations in all life forms, wholly in terms of 'for the good of the genes'. Dawkins's selfish-gene illusion stands or falls on the workability of this genic level of natural selection, the perpetual motor of evolutionary change in its relentless drive for its own self-propagation.

Where Did Dawkins Go Wrong?

To justify the concept of the gene as a unit of selection, Dawkins builds upon a number of erroneous assumptions. I will begin with two of the central assumptions to the selfish-gene illusion.

The first assumption states that what is important in evolution is the survival of the 'optimon' (a term coined by Dawkins), defined as that unit which benefits from a naturally selected adaptation. Such a unit must be an immortal, self-replicating entity and the gene (and not the organism) is the only true immortal, self-replicating entity. From this basis Dawkins concludes 'the fact that genes in any one generation inhabit individual bodies can almost be forgotten'.[3] Hence, Dawkins's logic produces the statement: 'elephant DNA is a gigantic program which says "Duplicate Me" by the roundabout route of building an elephant first'. The elephant is a digression. Within the bounds of this logic, Dawkins's metaphor of the river out of Eden is apt: individuals represent transient and stationary collections of genes which ultimately disperse and regroup in future downstream individuals as a consequence of the sexual shuffling of genes between generations. Only the genes are eternally flowing and self-replicating; only the genes are the units of selection, for they directly control which of their robotic digressions go forth and multiply.

The second assumption concerns the nature of evolved biological structures. For Dawkins all structures are adaptations (defined as having arisen through the process of natural selection) and all are 'improbable perfections' that could only have arisen by natural selection. Hence, 'eyes are perfections of engineering – should any parts be rearranged then that would make them worse'. This view leads naturally to the opinion that 'each species is an island of workability set in a vast sea of conceivable arrangements most of which would, if they ever came into existence, die'. Organisms are tiny islands surrounded by an 'ocean of dead unworkability'. Beware the unnatural monsters lurking in the swamps beyond the natural tree of life – naturally born of natural selection.

The 'Paradox of the Organism'

There is from the outset an inherent tension, recognised by Dawkins, between these two assumptions – the first that the gene, in its self-replicating eternity, is the single, selfish unit of selection, and the second that complex biological structures and functions are 'improbable perfections' of natural engineering, only made possible through natural selection. (See Stephen Jay Gould's chapter for the limits of natural selection.) If the *raison d'être* of each and every gene is only to make more copies of itself, come hell or high water, then how can complex parts of individuals (eyes, hearts, photosynthesis, consciousness, etc.) come into being, requiring, as they do, an intimate, co-ordinated behaviour of tens of thousands of similar genes?

Dawkins's persistence with this self-imposed 'Paradox of the Organism', arising from his belief that the organism should by rights 'be torn apart by its competing replicators', leads him inexorably to the ultimate conceit: 'the organism functions as such a convincingly unified whole that biologists in general have not seen that there is a paradox at all!' So there we have it: 'the logical correctness of the principle is inescapable'.

Yes, We Have No Paradox

Dawkins's 'solution' to the 'paradox' of his own making is that genes subdue their own individual selfish pursuits by collectively deciding on a shared list of *desiderata* (Dawkins's coinage) which ensures that they all end up in bodies producing successfully functioning gametes. To quote: 'They all "agree" over what is the optimum state of every aspect of the phenotype, all agree on the correct wing length, leg colour, clutch size, growth rate, and so on.'

At this point, I think it is legitimate to ask, are the genes selfish or not? Is each and every gene a unit of selection or is the cohesively functioning organism (the collective love-in of *desiderata* lists) the unit of selection? Technically, in order to exit from the cul-de-sac of his 'paradox', Dawkins has, consciously or not, been forced to redefine the gene in terms of transient and ephemeral phenotypic functions which, as I shall argue below, are the only true determinants of selection. For Dawkins, 'surviving genes are those that flourish when rubbing shoulders with successive samplings from the genes of the whole species, and this means genes that are good at co-operating'. Hence, the co-operative gene–gene *interaction* now becomes the inevitable focus of

Dawkins's scheme, and the interaction *is*, of course, the phenotype. By phenotype I mean everything beyond the gene starting with the protein products of the genes through the network of molecular interactions, leading to the all-singing, all-dancing, reproducing individual organism. The gene-as-organism has become Dawkins's roundabout route to retaining his selfish gene whilst solving the pseudo-paradox of the organism: having his cake and eating it. Is it any surprise, given the tortured and make-believe logic of all of this, 'that biologists in general have not seen that there is a paradox at all!'

Dawkins's selfish genery propagates a nonsense that is genetically misconceived, operationally impossible and seductively dangerous. It is Dawkins's dangerous idea, not Darwin's dangerous idea, which is seriously misleading. Theorists from diverse disciplines seem, unfortunately, quite happy to accept that evidence for a genetic contribution to complex human behavioural or morphological traits inevitably means evolution of that trait by a natural selection of selfish genes.

If we are to return the genes to their rightful place in biology we have to find a way out of this impasse.

Who Reproduces?

Outside the test tubes of the organic chemist, there is no such thing as a self-replicating molecule in biology and probably there never has been. Contemporary DNA relies for its replication on tens (maybe hundreds) of protein enzymes. In its turn, the correct synthesis of proteins is reliant on the coded information embedded in the combinatorial permutations of the four nucleotide bases of DNA and on multitudes of enzymes knee-deep in the process of translation. In all probability there has been an intimate co-existence and co-evolution between the different kinds of informational macromolecules (DNA and proteins) in the primeval soup. Recent findings that RNA can act as an enzyme have led to the erroneous belief that it can function as a self-replicating molecule. Not one such profligate molecule has emerged to date from quadrillions (10^{24}) of artificially synthesised, random RNA sequences.

The only free-standing biological entity capable of self-reproduction is the cell – the full-blown phenotype, part of whose diversity is governed by the genotype. The genotype cannot self-replicate; the phenotype can. For sexual organisms, whether they are single cells or multicellular, we need to add the caveat that reproduction is, by the very nature of the sexual process, neither exact nor involving only one side. It takes two to tango; but the resultant progeny, whilst unlike either parent, are obviously more like their parents than they are to any other parents. On this basis sexual organisms can be said to reproduce autonomously in pairs; genes cannot, singly or in pairs. The ability of each strand of the DNA double helix to be a template for the more-or-less correct synthesis of another partner strand relies wholly on the biological capacity of the phenotype to reproduce. DNA is not an eternally self-replicating, free-standing unit of selection, as strongly implied both in reality (the hard line) or metaphorically (the soft line) in Dawkins's view of evolution by natural selection.

Is Natural Selection a Process?

The selection phenomenon requires some living entity to reproduce (or over-reproduce, if we return to Darwin's original reliance on the Malthusian principle). This is achieved by the organism. (See Steven Rose's chapter.) The selection phenomenon also requires an inheritance of coded information from one cell to another or from one generation to another. This is achieved, to all intents and purposes, by the DNA. Dawkins's first error in this context is to suppose that, because there is a genetic contribution to the form, behaviour and reproduction of the phenotype and because this contribution is inherited, the gene is *ipso facto* the unit on which selection acts. The second error is to suppose that, because there is a genetic contribution to a given complex function, natural selection acting alone is the only mechanism capable of achieving the 'improbable'.

'Natural selection' is nothing more than the one-off, never-to-be-repeated, passive outcome of a unique set of interactions of newly established unique phenotypes with their local environment, at each generation. The genetic uniqueness of individuals is a consequence of mutations and the random shuffling and recombination of chromosomes during the sexual process. Naturally we need to add to this genetic uniqueness the unique environmental contribution to each individual's development. Hence, the differentials in reproductive success between such unique phenotypes are also unique in each and every generation. As a result of such never-to-be-repeated conditions, some genetic variants are passively and differentially passed on, in new combinations, to make a contribution to the next set of unique phenotypes. Phenotypes are 'selected' (or, more accurately, 'self-select' themselves) for greater reproductive success as a consequence of their unique sets of physical and behavioural attributes, as they exist in a given environmental milieu. As a consequence, genes are inevitably 'sorted' as they become redistributed into the next set of individuals.

Whatever phenotypic selection and genetic sorting might have happened at any given generation, it is the uniqueness of the components involved with the selection phenomenon which ensures that the same selections and sortings will never reoccur. On this basis, natural selection is not only not an *active* process (that is, some force directly selecting phenotypes or, if you will, selecting Dawkins's genes), it is not a *process*, as such. Darwin himself advised against the potential confusion over this issue, advice ignored by generations of so-called Darwinists, culminating in Dawkins. Hence, it is illegitimate to give quantitative numbers (coefficients) to genes reflecting their genic 'power' (Dawkins) to influence the results of selection. That such coefficients have been used over the past eighty years in order to model natural selection does not mean that this is an adequate representation of what is happening in nature; it is merely a simplifying device to make the mathematics more tractable. Organisms evolve under very complex internal and external conditions which have not so far been captured by the available models.

The idea that what is important is a gene's average effect on the reproductive successes of its bearers, generation after generation, falls into the difficulty that evolution could not proceed by taking each gene, one at a time, whilst holding constant the effects of all the other supposedly selfish 99,999 genes which make up a typical 100,000 genes' worth of contribution to an advanced primate phenotype. Were this to be done for each gene in turn, then there would not be enough time, since the cooling of the earth, to have evolved what we know has evolved.

One way out of this problem is to follow the same approach used to overcome the pseudo-paradox of the organism. That is, not to consider selection as acting on each separate gene (whilst keeping the rest of the genetic contribution effectively average and neutral), but instead to consider selection as working on several genes at a time. Once again we need to consider the phenotypic consequences of the ways genes interact during development. Much of the recent excitement over the genetic basis of development in animals and plants concerns the hierarchies of gene–gene interactions, often involving the genetic circuitry that controls the on-off expression of genes, as development unfolds. A given gene can be involved in a number of unrelated gene control circuits and a given circuit can involve a number of different controlling genes.

Long Live the Ephemeral Phenotype!

It would be simpler and more logically coherent to abandon the idea altogether of the gene as the unit of a process of selection and to focus instead on the role of the ephemeral phenotype; that is, the self-reproducing product of multilayered and motley genetic interactions. There are no genes for interactions as such: rather, each unique set of inherited genes contributes interactively to one unique phenotype. Hence, it is the very transient nature of unique phenotypes, coupled to their biological trick of knowing how to reproduce, which makes them, through their specific exchanges with the local environment, the only determinants of the operation of selection. Selection is here and now, not everywhere for all time.

If we abandon the idea of selection as an active process requiring constant selection coefficients attached to single genes, or even to collections of independent single genes, then we no longer need to assume that only a self-replicating entity can be the 'optimon' unit of selection. Nor is it eternally flowing down the river out of Eden. Indeed, there is no need for a *unit* of selection, whether free-flowing or trapped on a sunken branch. Selection can be pushed any which way, in any generation: we cannot make predictions from what happened in one generation as to what might happen at the next because of the sheer variety of unique phenotypic determinants generated by sex at every turn. A shifting population of organisms is totally different in structure from a constant population of water molecules, which can be predicted to behave in ways prescribed by known physical laws.

This is not to suggest that a trend, say, towards longer necks of giraffes was ever a one-off outcome of differentials in reproductive success of phenotypes in a single generation. As Darwin and many others have proposed, the longer-necked *phenotypes* at any given generation are the result of the more reproductively successful longer-necked *phenotypes* at the previous generation, having previously passed on the genes which contributed to longer necks. The outcome of natural selection, even as a one-off comparison of reproductive successes amongst unique individuals in any given generation, is contingent upon the one-off comparisons in all previous generations. This line of apparent regression does not of course mean that we should replace the complexity of selection by imaginary self-replicating genetic units with fixed selective 'powers'.

No Genetic Blueprints

The species-specificity of the form and behaviour of a new developing individual is not specified in the DNA. There is no blueprint, or set of instructions, in the DNA saying, 'Make me an elephant – I feel like digressing.' Genes are wholly 'ignorant' of their effects on development. The final, multicellular individual, bristling with Darwinian adaptations and much else, is the product of piecemeal interactions of proteins with proteins, proteins with DNA, proteins with RNA, and RNA with DNA. The interactions are local and blind to everything that has gone on before and everything that has yet to come during development. Ontogeny is, in part, a progressive unfolding of the expression of genes in particular cells and at particular times. Gene expression itself is under the control of the protein products (activators and repressors) of other genes. Hence, the specific pattern of unfolding which ensures that each new human fertilised egg usually develops into a multicellular, reproducing human adult is intimately bound up with the evolutionary history of the human genome. As with everything else, this has been sorted as the after-effect of billions of instances of differential successes in reproduction amongst unique and ephemeral phenotypes, along with molecular drive and genetic drift (see below), stretching back in time to the appearance of the first DNA-protein interdependent cell.

The 'New Genetics' and the 'New Evolution'

Scattered in the voluminous writings of Dawkins, there are sufficient statements to support most of what I've said here and to make anyone wonder what the fuss is all about. Are there indeed any substantive disagreements? Dawkins can often be seen to face both ways (the selfish gene: now we see it, now we don't). Dawkins's hard line is that he has opened our eyes to a dramatic new way of thinking (what he terms a 'transfiguration') about the genetic motor of natural selection; his soft line is that he is saying nothing new of major importance, for, yes, deep down, he recognises that it is the phenotype which is the prime mover, as Darwin rightly conceived. This ability to look both ways at once was likened by Dawkins himself to the visual illusion of the Necker Cube presenting two simultaneous orientations of a 3-D cube from a 2-D paper image. We should not be taken in by this edgy ambivalence; the perceived thrust of Dawkins's writing is about only one thing: everything of functional importance and complexity is an adaptation fashioned by natural selection working for the good of selfish replicators. The caveats, the qualifications, the 'ifs' and 'buts', are not part of the grand illusion.

The hollowness of Dawkins's arguments is revealed by what he fails to discuss as important for evolutionary theory drawing from the heaped treasure trove of late twentieth-century biology, even though some of these findings are several decades old. The 'new genetics' and the ensuing 'new evolution' revolve around features of modularity, redundancy, combinatorial permutations, molecular drive and molecular co-evolution which collectively place the organism and its functional evolution well out of the reach of Dawkins's selfish 'optimons'.

Modularity

The modular nature of genes, proteins, developmental pathways and organs is one of the far-reaching new surprises of biology. (Caution! This concept of modularity has little relationship to the concept of modularity as used in evolutionary psychology; see Annette Karmiloff-Smith and Barbara Herrnstein Smith's chapters.) A module is an independent unit of function that can interact, in a variety of permutations, with other modules, seemingly without regard for the overall form and behaviour of an organism. At the DNA and protein levels a module consists of a block of nucleotides or amino acids, shorter than the gene, with a prescribed function. Perhaps surprisingly, modules can be shared by totally unrelated genes and good estimates of the total number of gene or protein modules amongst the, say, 100,000 different genes and proteins across life's forms yield a figure of between 1,500 and 2,000 modules only. This means that genes and proteins differ primarily by the specific permutations of modules, both with regards to module type and to the number of copies of each type. In other words, a gene or protein can be structurally defined by a given subset of modules, many of which can occur more than once in a given gene. The relatively small number of different modules extensively shared by unrelated genes and their regulatory regions uncover the existence of universal genomic mechanisms of rearrangement.

The Fluid (Non-Mendelian) Genome and Its Evolutionary Implications

To cut a long story short, the past three decades of research have demonstrated that genomes of all forms of life are in a constant state of flux due to a variety of mechanisms of DNA turnover, two of which are 'transposition' and 'slippage'. The first of these is responsible for the ability of modules to jump from one gene to another, or from one regulatory region to another. The second accounts for the fluctuations in the numbers of copies of modules in any given genetic region. Another mechanism of flux involves the alteration of the genetic constitution of one gene to resemble the genetic constitution of an alternative gene variant by 'gene conversion'. The ability of two slightly different gene variants (*A* and *a*) to become *AA* or *aa* is due to the two relevant double helices of DNA breaking up and reconnecting in different ways. There are half a dozen other bizarre features of DNA instability which I have no space to describe.

The important feature of all such mechanisms is that they are *non-Mendelian*: in other words, the numbers of copies of, say, *A* or *a* can go up or down in the lifeline of an individual and ultimately spread through a sexual population in a way that is operationally distinct from the sorting of gene frequency changes associated with natural selection amongst phenotypes. Genes that obey the original rules of segregation discovered by Mendel do not increase or decrease in number unless forced to do so by, say, natural selection. The non-Mendelian mechanisms ensure that gene mutations can alter in number under their own steam.

The original Mendelian laws of inheritance, based on the random behaviour of chromosomes during the sexual process, are assumed to give rise to populations that are in genetic equilibrium. In other words the laws of inheritance cannot, *in themselves*, bring about a change in the average genetic composition of a population (that is, they cannot ensure what Darwin called 'a modification by descent'). This realisation gave the acceptance of natural selection a much-needed boost in the 1930s: the marriage of Darwin with Mendel (the so-called modern synthesis) meant that observed evolutionary changes could be accepted universally as having happened through the phenomenon of adaptation sorting through well-behaved Mendelian genes. However, the discovery of a wide range of slowly acting, but persistent, non-Mendelian genomic mechanisms underlined the general instability of DNA and the potential to spread ('molecular drive') new genetic variants through sexual populations, with the passing of the generations.[4]

The genome is a complex mix of the products of turnover whose mechanisms operate at different rates, with different biases and on different unit lengths of DNA, simultaneously within a given region of DNA. Detailed analysis of hundreds of genes and their regulatory regions shows that no section of DNA is refractory to the operation of one or other turnover mechanism. The genome is riddled with multiple copies of jumping and moving bits of DNA, between and within genes.

New genes and regulatory regions can be created, therefore, from floating modular non-Mendelian units that come together into new genetic permutations. The non-Mendelian manner in which new permutations come into being also influences the representation of such newly formed permutations in the population. The mechanism of spread is simple: if a module makes an extra copy of itself, say, through 'gene conversion', then, at conception, each copy is deposited in a different individual in the next generation. In each of these new individuals 'gene conversion' can occur again, so creating yet more copies. Through the combined efforts of turnover and sexual reproduction, a new genetic element can occupy an increasing number of individuals over the generations.

With this internally driven process we break out of the axiom of Darwinian–Mendelian evolution, which assumes that all mutations are static, one-off occurrences waiting for selection to promote or demote at the level of the population. Many novel genetic constructs can spread, slowly but inexorably, through sexual populations as a consequence of the ubiquitous genomic mechanisms of turnover.

Teaching Old Modules New Tricks: 1+1=7

Modularity is also a key feature of developmental processes and organ production, which rely on local, combinatorial interactions that proceed without regard to the organism as a whole.[5] For example, legs and antennae in insects are completely interchangeable modular constructs under the control of modularly constructed genes – called homeotic genes. Similarly, experimental manipulation of other modular genes can generate perfectly formed insect eyes in legs and wings of insects, even though the manipulated 'eye' genes were derived from mammals.

The movement of modules around the genome from gene to gene and from regulatory region to regulatory region ensures that new constructions can engage in a whole variety of new functions. Biology has found an easy way of teaching new tricks to old modules through the expediency of sharing new modular neighbours. Genes are pleiotropic: that is, they are capable of teaming up with many other genes to generate complex phenotypic functions. They can be involved in many different types of cellular and developmental processes. It is highly unlikely that there are consciousness-specific or language-specific genetic modules in humans that are not already in use in a multitude of other cellular processes throughout development. No one gene can selfishly self-select itself without precipitating a multitude of developmental problems, given its multitude of activities.

The evolutionary implications of modularity are enormous, notwithstanding Dawkins's statement 'that evolution is unlikely to incorporate homeotic mutations'. Incorporation of regulatory changes by such genes and many others is precisely what evolution has been doing since the Cambrian. The modular constituents of genes, proteins and many developmental operations, shared across widely divergent taxa, are akin to giant Lego sets. Many supposedly complex adaptations are not the end result of the addition of hundreds of brand-new genes; rather, they are the products of new combinatorial permutations of a limited set of free-floating and widely distributed non-Mendelian modules.

With this knowledge, we can dispense with Dawkins's belief that complex functions are 'improbable perfections' arrived at only as naturally selected adaptations. The Lego construction set can get as big as we wish, starting from a limited set of building blocks. The 'improbability' of achieving something that is perfect and reproduces does not increase as evolution progresses. Complexity is a combinatorial phenomenon. In biology 1+1=7 and the argument for selection based on the claim of improbable 'complexity' is unwarranted. There is no need to follow Dawkins into the mire of the 'fallacy of perfection'.

In order to make these arguments clear, I need to describe the ways in which we are beginning to understand the close relationships between natural selection and the molecular turbulence within populations of genomes. There is as much a problem of the internal 'tangled bank' of the genome to which selection needs to provide a solution as there is of Darwin's external 'tangled bank' of the ecology.

Locks and Keys

One prevalent analogy for natural selection is that it provides a solution – a key – to a given problem – a lock. Imagine two levels of locks and keys: one between the organism and its niche and the other between one molecule and its interactive partner. Although a niche can really only be usefully defined in terms of the dynamic interaction between the choosing organism and its environment (see Pat Bateson's chapter), it is generally believed that selection is a process which drags organisms, kicking and screaming, to an established niche (up the dreaded slopes of Mount Improbable). Leaves grow freely on the tops of trees and giraffes' necks were optimised (adapted) by selection to reach them. The same honing and optimisation is considered to have happened in terms of intermolecular interactions. In modern-day story-telling we are free to imagine any ecological 'lock' in the distant days of hunting and gathering by our proto-human ancestors to which the 'key' of contemporary behaviour (sexuality, aggression, fear, parental invest-ment, etc., etc.) has been selective fashioned. For example, it is claimed that stepchildren have a rough time at the hands of their stepfathers because of the evolved genetic imperatives of kin selection and not because of the enforced tensions inherent in small nuclear second families in contemporary society. (See Hilary Rose's chapter.)

The underlying image of selection is that there was always sufficient genetic variation for the current adaptive solutions to have evolved. Furthermore, it is assumed that the emergent biological functions are naturally engineered by natural selection. No other functional and reproducing entities, within the assumed strictures of locks and keys, could have successfully evolved. Out there in the largely unoccupied space of phenotypes lurk only the unnatural monsters of legend. Taking the case of snail shells as his phenotypic example, Dawkins likens this outer darkness of unoccupied space to rows upon rows of empty shelves in his n-dimensional museum of all possible shapes of snail shells. For Dawkins this vast unoccupancy is due to mutant shapes which were of no use to selection in its quest to produce robotic miracles of biological engineering, with their alleged sole purpose of propagating the genes. Again, within the confines of hindsight of life *as we know it*, Dawkins can state, 'whenever in nature there is a sufficiently powerful illusion of good design for some purpose, natural selection is the only known mechanism that can account for it'. Indeed, this process of natural selection is deemed by Dawkins as sufficiently powerful to produce organisms whose 'evolvability' is so efficient that 'they save natural selection from wasting its time exploring vast regions of search space *which are never going to be any good anyway*' (my emphasis). Search space means the shelves in Dawkins's museum.

We Are All Monsters Now

Developing a counter-argument to this 'fallacy of perfection' does not entail considering whether life as we know it is a good, natural design, produced by phenotypic self-selection, generation after generation. Instead, we should ask what living organisms might have been like had species diversity taken different routes, at each and every instance at which novelty was produced. There is only one tree of life; it happens to occupy, in the limited time available since the origin of life, only a minute fraction of the n-dimensional space of genotypes/phenotypes. It is illegitimate to argue that the supposed harmoniously functioning living 'designs' *as we now see them* are the 'improbable perfections' engineered solely by non-random selection. Because there is only one tree of life it is illegitimate to assign probabilities to single events. This statistical failure is particularly pertinent to the claimed one-off, passive nature of selection.

For example, we can describe how the evolution of the human eye took a known series of steps from the earliest light-sensitive molecules on the surface of living cells to its full-blown 3-D Technicolor glory. We can also say, for the sake of simplicity, that at each of the steps there were only two choices: the right choice, *as defined by what we now know was the right choice*, and the wrong choice, the evolutionary path never explored. However, armed with this *post hoc* knowledge, it would nonetheless be a mistake to follow Dawkins's implicit logic. Calculating that the chance of making all the right choices, at each and every step and for all steps in the *known* sequence of events, is one in two to the power of the total number of steps. It is this procedure that leads to the fallacy of the 'improbability' of stringing together the correct sequence of events by chance, and which in turn leads to the belief that only selection can bring about the 'improbable'. This miscalculation of Dawkins, looking vertically through time at the 'known' evolution of the eye, echoes the well-known miscalculation made by the astronomer Fred Hoyle, looking horizontally at the overall sequence of amino acids which make up a typical protein. Famously, for Hoyle, the spontaneous formation of a typical protein is equivalent to a hurricane throwing together a Boeing 747 from a junkyard.

Interestingly, Dawkins recognises that even the arrangement of junk in a *particular* junkyard is as improbable as the Boeing 747 arrangement. Hence, in order to rescue the Boeing 747 as being functionally special, the product of non-random selection, he needs to assume it is an improbable 'perfection' superior to all other arrangements. However, the simile is badly chosen when we consider biological functions, and Dawkins's conclusion that all the current wonders of the eye would never have been achieved by 'almost all random scramblings of the parts' has the benefit of hindsight. With this he is right to say that 'there is something very special about the *particular* arrangement [of the eye] that exists'. But he is wrong to say that the huge majority of all other *particular* arrangements, which he admits are equally as improbable, would not produce useful devices for seeing – that would only be true if we were after achieving the human eye as we know it.

Not Improbable and Not Perfect

Even were we to assume that natural selection is the only show in town, we need to ask what would have happened if, say, at some intermediate step in the evolution of the eye, the 'wrong' choice was made ('wrong' with reference to what we know did happen). Let us suppose that this 'wrongness' resulted in an eye on an insect's leg or antennae, as can be achieved experimentally through manipulation of the genes involved with eye development. *With hindsight*, we know that this would have scuppered the observed, real history of the eye that ended up with it on the head. *With hindsight*, we can classify such an eye-on-a-leg phenotype as monstrous and unnatural – ready for the chop by the relentless perfectionism of selection. *However*, and this is a very large qualification, we cannot make any predictions from the comfort of *life-as-we-see-it-today*, of what life might have been like had eyes-on-legs occurred. There might very well have been some selective advantage to eyes on other parts of the body upon which all subsequent contingent events would have flowed – leading to contemporary life forms with eyes everywhere but their heads – equally as wondrous 'improbable perfections' as eyes-on-our-heads. The same could be said about the diverse range of functions of the eye: for example, the evolution of an eye capable of using a spectrum of electromagnetic radiation that allowed us to see in the dark, rather than perceive 'visible' light. This would not have led to life *as we know it*, but it would have led to some probable function – not 'improbable' and not needing to be defined as 'perfect'. There is indeed an anaesthetic of familiarity in the way we argue with hindsight that makes a laughing stock of all the supposed waste-of-time monsters occupying the unfilled phenotypic space.

What can be said about eyes applies equally as well to many of the basic modular components of our internal cellular and developmental machinery – the genetic code, protein assembly, membrane transport, energetics, muscles, nerves, and so on. In addition to this, there are alternative means of producing useful novelties distinct from natural selection yet which interact with it. These alternative non-selective, but *not* anti-Darwinian, means of evolution signify a relaxed tolerance between organisms and their environment and between functionally interactive molecules. This is far removed from the tight specificity and essentially intolerant perfectionism of the lock-and-key imagery of selection.

Exaptations: Functions for Nothing

Stephen Jay Gould and Elizabeth Vrba have coined a new word, 'exaptation', to cover established biological functions that did not arise as Darwinian adaptations.[6] My favourite exaptive process arises from the neutral theory of evolution. Neutral theory[7] predicts that mutations in sequences of DNA which have no, or very little, consequence for reproduction may spread in a population by chance given the inherent instabilities in sizes of populations with time. Such mutant sequences are not prevented from spreading by selection for they have no important effects on phenotypes.

It is conceivable, however, that once a neutral mutant gene has spread to all members of a population there might be subsequent changes in the environment which could elicit a new function from the gene in question. This new function would be an exaptation as it did not arise initially through the differential selection of phenotypes. Driving on the left in the UK could be understood as an exaptation based on the spread of a historical decision. Driving a car on the wrong side today and experiencing a dramatic loss of fitness does not signify that the spread of the rule depended on the selective weeding out of those who didn't obey the rule and lost the capacity to reproduce.

Molecular Co-evolution, Adoptation and the Dynamics of Tolerance

Tolerance arises out of two features: genetic and functional redundancy, and the particular cohesive dynamics of change at the population level as a consequence of the internal mechanisms of DNA turnover.

When genomic turnover gives rise to several identical or near-identical genes producing similar proteins, then the organism can be buffered against the effects of mutation in one of the genes. For example, genes coding for the family of globin proteins arose by a series of duplications followed by divergence. One copy of a gene can retain its original function, whilst another mutant copy might get involved with another function. The extent of buffering depends on the number of genes and their effects on phenotype. Genetic buffering also occurs amongst the repetitive modules that can make up a gene or its regulatory regions. The phenotype can also be buffered against the mutation effects of the genes involved with a developmental process, if there are (as is often the case) two processes with the same end result.

Internal buffering and such a slow cohesive change in the genetic constitution of a population provide the context in which natural selection can occur. However, if selection does not like what's going on, the phenotypes with too many copies of the variant unit will be of less reproductive value. However, compensatory mutations in other genes can also occur. In this way, the interactive function is maintained, or slightly modified, and the two interacting partners can be said to have co-evolved. The ability of functionally integrated genetic systems to co-evolve is an important route for the evolution of novel functions without a complete breakdown in services. For example, we can envisage evolution having learned the trick of permitting ongoing changes in the design of an aeroplane, whilst the bloody thing is flying in the air! Biological evolution does not need to have the plane crash every time one of its thousands of supposedly 'selfish' parts (genes) decides to mutate.

Subtle changes in the ways biological functions take place and the establishment of species-specific novelties are attendant on the phenomenon of 'molecular co-evolution' – the continuous, friendly conversation between molecular partners – all of which is enacted through the coherent phenotype. The phenotype is not a paradox; it can be sensibly understood as a functioning, reproducing unit both in terms of its development and its evolution.

When populations are evolving cohesively and molecular co-evolution is occurring it is reasonable to suggest that such a cohesively changing population can as well adopt some previously inaccessible component of the environment (the so-called niche) as it can be adapted to the environment individually by natural selection. The inherent flux and redundancy of the genetic material, coupled to the widespread sharing of the modular constructs that underpin diverse biological functions, ensures that there is a much looser relationship between organisms and their environment than was supposed when there was only natural selection as the motor of evolutionary change.[8]

There is a very large role for natural selection in coping with the internal 'tangled bank' of the genome. However, there is far more to life than natural selection. Biological functions are a complex mix of adaptation, exaptation, molecular co-evolution and adoption arising from the properties of turbulent genomes in turbulent environments (see Figure 1). The forces that underlie evolution are tractable but not reducible to the supposed primacy of the 'masters of the universe' – the so-called selfish, eternally self-replicating Mendelian genes.[9]

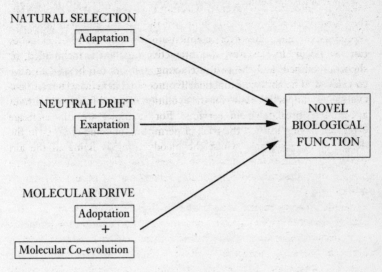

Figure 1 Three Activities: Three Outcomes: One Product

References

1 E. Sober, and D. S. Wilson, *Unto Others: The Evolution and Psychology of Unselfish Behavior* (Harvard, Harvard University Press, 1998).
2 W. D. Hamilton, 'Altruism and Related Phenomena, Mainly in Social Insects', *Annual Review of Ecological Systems*, 3 (1972), pp. 193–232.
3 All direct citations are taken from R. Dawkins, *The Selfish Gene* (Oxford, Oxford University Press, 1976); *Climbing Mount Improbable* (Harmondsworth, Penguin, 1997); and 'Parasites, Desiderata Lists and the Paradox of the Organism', *Parasitology*, 100 (1990), pp. 63–73.
4 G. A. Dover, 'Molecular Drive: A Cohesive Mode of Species Evolution', *Nature*, 299 (1982), pp. 111–17; G. A. Dover, 'Molecular Drive in Multigene Families: How Biological Novelties Arise, Spread and Are Assimilated', *Trends in Genetics*, 2 (1986), pp. 159–65; G. A. Dover, 'Evolving the Improbable', *Trends in Ecology and Evolution*, 3 (1988), pp. 81–4.

5 R. Raff, *The Shape of Life: Genes, Development and the Evolution of Animal Form* (Chicago, Chicago University Press, 1996).

6 S. J. Gould and E. Vrba, 'Exaptation – a Missing Term in the Science of Form', *Paleobiology*, 8, pp. 4–15; S. J. Gould and R. C. Lewontin, 'The Spandrels of San Marco and the Panglossian Paradigm: A Critique of the Adaptationist Programme', *Proceedings of the Royal Society*, 205 (1978), pp. 581–98.

7 M. Kimura, *The Neutral Theory of Molecular Evolution* (Cambridge, Cambridge University Press, 1983).

8 G. A. Dover, 'Observing Development Through Evolutionary Eyes: A Practical Approach to Molecular Co-evolution', *Bioessays* (Special issue: *Evolution and Development*), 14 (1992), pp. 281–7; G. A. Dover, 'The Evolution of Genetic Redundancy for Advanced Players', *Current Opinions in Genetics and Development*, 3 (1993), pp. 902–10.

9 The argument of this chapter is expanded in my book *Dear Mr Darwin: Letters on the Evolution of Life and Human Nature* (London, Weidenfeld & Nicolson, 2000). I am grateful for the support of a Leverhulme Trust Fellowship.

5

Why Memes?

Mary Midgley

Thought Is Not Granular

What does it mean to understand the workings of human thought or of human culture? What kind of understanding do we need here? Is it the kind that might be achieved by atomising thought – by analysing it into its ultimate particles and then connecting them up again? Or is it, rather, the kind of understanding that we normally mean when we speak of understanding a suggestion, or a policy, or a word by placing it in a context that makes it intelligible – by supplying an appropriate background?

In general, of course, these two patterns are not alternatives. Normally, we use both together as complementary, helping us both to understand and explain. Both in science and in everyday life, we reach equally readily for either tool as the case requires. Of late, however, it has been strongly suggested that, in studying thought, the atomising approach is the only truly scientific one and should take precedence over other methods. Accordingly, Richard Dawkins has suggested that the scientific approach to culture is to split it into units called *memes*, which are in some ways parallel to its atoms, in others to its genes, and to study their interactions.[1] This proposal is an entirely understandable move in view of the success of similar methods in the physical sciences. It is always natural to hope that a method which succeeds in one area will help us in another and, in this case, there have been strong and respected lines of thought in our tradition which have demanded just such a project.

All the same, these lines of thought cannot really help us in this quite different situation. The trouble is that thought and culture are not the sorts of thing that can have distinct units at all. They do not have a granular structure for the same reason that ocean currents do not have one – namely, because they are not stuffs but patterns.

There is nothing mystical or superstitious about this. Sea water is indeed a stuff or substance with units. It can be divided – not, indeed, into the hard indivisible little grains that Renaissance physicists expected, but still into distinct, lasting physical particles. By contrast, the currents themselves are patterns of movement – ways in which this water flows – and they form part of a wider system of such patterns which surrounds them. To understand the currents one must first investigate these wider patterns. Of course the microstructure of the water itself is sometimes relevant here, but usually it is just a standing background condition. The microscope is not the first tool that scientists reach for when they want to understand the distribution of sewage in the oceans or – even more obviously – when they are analysing patterns of traffic flow. The first movement of understanding in such cases has to be outward, to grasp what is happening in the context. But thought and culture too are moving and developing patterns in human behaviour, ways in which people think, feel, and act, not entities distinct from those people. Since such patterns are not composed of distinct and lasting units at all, it is not much use trying to understand them by tracing reproductive interactions among those units.

Incidentally, the name 'meme' itself is of some interest. Dawkins explains that he abbreviated it from *mimeme*, meaning a unit of imitation, and both words are evidently modelled on *phoneme*, which is a term invented by linguists early in the twentieth century to describe a unit of sound. 'Phoneme' however quickly turned out to be the name of a problem rather than of a fixed ultimate unit. As John Lyons, Professor of General Linguistics at the University of Edinburgh, put it in 1970,

> Most linguists, until recently at least, have looked upon the *phoneme* as one of the basic units of language. But they have not all defined phonemes in the same way (and have frequently arrived at conflicting analyses of the same data). Some linguists have described phonemes in purely 'physical' terms; others have preferred a psychological definition. Some have argued that grammatical considerations are irrelevant in phonological analysis; others have maintained that they are essential. These are among the issues that have divided the various schools of phonology in recent years . . .[2]

And the problem has certainly not got any simpler since he wrote. In short, the sound of speech as we hear it turns out not to be granular, not to have definite units. It is a continuum that can be divided up in various ways for various purposes. The original hope of atomising it seems to have flowed from a general confidence in atomising which was rather prevalent at that epoch. One might compare the mechanistic biologist Jacques Loeb's notorious concept, advanced at the beginning of the twentieth century, of tropisms as an ultimate unit that would explain all motion in plants and animals.[3] In most fields this approach has not turned out to be useful and there seems no clear reason to revive it today.

The Hope of Unification

The meme project has, however, been rather widely accepted because it is exactly what our Western cultural tradition has been waiting for. For two centuries, admirers of the physical sciences have wanted somehow to extend scientific methods over the whole field of thought and culture. They wanted it for the entirely proper reason that they wanted to reunify our thinking, to heal the breach in our world-view made by Descartes's division between mind and matter, between the physical sciences and humanistic ways of thinking. The methods used have not, however, gone half deep enough. What was needed was genuinely to abandon the dualistic approach – to stop considering a human being as a hybrid and treat it as a whole that can properly be examined from many different aspects. Instead of doing this, theorists have tended still to assume that they were dealing with two separate, parallel kinds of stuff or substance, namely, mind and matter. This residual dualism makes it look as if we can unify the two by just extending the methods which we have used so successfully on matter to the parallel case of the other stuff – mind. It seems to promise that in this way we can fulfil the positivist programme, proclaimed by Auguste Comte, that human thought should progress steadily away from religion through metaphysics to a triumphant terminus in science.

That programme, however, is not really an intelligible one. It only looks plausible because of an ambiguity in the idea of science. As the historian C. B. A. Behrens puts it:

The [French] Philosophes believed that enlightenment had been vouchsafed to them by the discoveries of the seventeenth century, particularly those of Newton, which had illuminated the nature of the physical universe, and those of Locke, which had done the same for the mind . . . Their ultimate purpose was to spread the belief that human behaviour, like the material universe, was amenable to scientific investigation, and that science and government should be studied scientifically in the interests of human happiness.[4]

But at that time the meaning attached to studying things scientifically was much wider than it is today, as is plain from the bracketing of Locke with Newton. Centrally, the term scientific still had the very general meaning of thinking things out for oneself in an appropriate manner rather than just relying on tradition or authority. As time went on, the rising success of the physical sciences gradually biased its meaning towards presenting them as the only model. And the opposition that some Enlightenment prophets proclaimed between science and religion, casting religion as the representative both of tradition and of political oppression, intensified this bias. Newton's example, too, tended to be used as a justification for any simplification on social subjects, as if the physical sciences always proceeded by making things simpler. Thus David Hume justified his reduction of all human motives to utility by remarking:

> It is entirely agreeable to the rules of philosophy, and even of common reason; where any principle has been found to have a great force and energy in one instance, to ascribe to it a like energy in similar instance. *This indeed is Newton's chief rule of philosophising* . . . Why do philosophers infer, with the greatest certainty, that the moon is kept in its orbit by the same force of gravity, that makes bodies fall near the surface of the earth, but because these effects are, upon computation, found similar and equal? And must not this argument bring as strong conviction, in moral as in natural disquisitions?[5] (Enquiry, 163 and 192. Emphasis mine.)

But not all simplification is scientific. Newton's greatness did not lie merely in simplifying the scene but in the prior work that enabled him to see which generalisation to back, which simple system to design. Science progresses just as often by making distinctions as it does by abandoning them.

Hume, then, was being misled by a surface likeness, imitating a superficial form of thought rather than penetrating to its point. But he was not alone in this. Many theorists during the later Enlightenment were ambitious to become the Newton either of psychology or of political science. They claimed scientific status for an almost infinite variety of simplifications pursued from various ideological angles, so that eventually the excesses of allegedly scientific prophets such as Marx, Freud and Skinner began to cause a great deal of alarm. This is why, in the mid-twentieth century, serious admirers of science, led by

Karl Popper, narrowed the meaning of the term *science* in a way designed to cover only the physical sciences themselves.

The Dream of Standardisation

This was a natural move, but it raises a difficulty which has not yet been properly faced about the status of other kinds of thinking. Though Popper's campaign was aimed primarily against ideologists such as Marx and Freud, on the face of things it also disqualifies the social sciences and humanities from counting as fully 'scientific'. And since the term *scientific* remains a general name for academic excellence, people conclude that these cannot be serious, disciplined ways of thinking at all. Social scientists and humanists therefore often feel that they ought to make their reasonings look as like physical science as possible. This is the demand that memetics satisfies.

We need to be clear why this whole ambition is unnecessary as well as impossible. The right way to remedy the Cartesian split is not for one half of the intellectual world to swallow the other but to avoid making that split in the first place. A human being is not a loosely joined combination of two radically different elements, but a single item – a whole person. We do not, therefore, have to divide the various ways in which we think about that person into two rival camps. These various ways of thinking are like a set of complementary tools on a workbench or a set of remedies to be used for different diseases. *Their variety is the variety of our needs.* The forms of thought that we need for understanding difficult social dilemmas are distinct from those that we need for chemistry and those again from historical thinking, because the questions that we must ask in these areas are of different kinds, though of course all these forms and all these questions are related in the context of life as a whole. Naturally, then, those forms have different standards of validity.

These different kinds of need for thought are actually just as diverse as our various physical diseases are. In medicine today the idea of a universal patent medicine such as was advertised in Victorian times, a remedy potent to cure equally colds, smallpox, rheumatism and cancer, would not seem plausible. Nor are supposedly universal tools much welcomed on the workbench. The sort of unity that thought actually needs is not the formal kind that the philosopher Daniel Dennett tries to impose in his book *Darwin's Dangerous Idea* by inflating Darwinism into a universal system.[6] (See Stephen Jay Gould's chapter for a detailed critique of Dennett.) It is a unity that flows adequately from the fact that we are studying a single world – the one that we live in – and that our thought arises from a single source – namely, our joint attempt to live in that world. The fact that all our ways of thinking deal with that world, which is a single world though a very complex one, unifies our thought sufficiently, just as the science of medicine is sufficiently unified by the fact that all its branches deal with the human body.[7]

Is any stricter, more formal kind of unity possible? The great rationalist thinkers of the seventeenth century were obsessed by the ambition to drill all thought into a single formal system. Descartes himself, as well as Spinoza and Leibniz, tried inexhaustibly to mend the mind–body gap by building abstract metaphysical systems powered by arguments akin to their favoured models of thought – logic and mathematics. They were answered, however, by empiricists such as Locke and Hume, who pointed out how disastrously this project ignores the huge element of contingency that pervades all experience. Since we are not terms in an abstract calculation but real concrete beings, we do not live in a pure world of necessary connections. We deal with this pervasive contingency by ways of thinking – such as historical method – which provide crucial forms for our understanding of the world, but which cannot be reduced to a single form.

Although both rationalists and empiricists made strenuous efforts to claim a monopoly for their own chosen forms of thinking, it has become clear, from Kant's time onward, that the map of thought must allow for a wider variety of methods. That map is far more radically complex than the seventeenth century supposed. It cannot be drilled to show a single empire. Dennett, however, persistently tries to square this well-known circle by imposing uniformity. He describes Darwin's 'dangerous idea' – that is, the idea of development by natural selection – as a 'universal acid . . . it eats through just about every traditional concept and leaves in its wake a revolutionised world-view, with most of the old land-marks still recognisable, but transformed in fundamental ways' (p. 63). This is, however, evidently a selective acid, trained to eat only other people's views while leaving Dennett's own ambitious project untouched:

> Darwin's dangerous idea is reductionism incarnate, promising to unite and explain just about everything in one magnificent vision. Its being the idea of an algorithmic process makes it all the more powerful, since the substrate neutrality it thereby possesses permits us to consider its application to just about anything (p. 82) [including] all the achievements of human culture – language, art, religion, ethics, science itself. (p. 144)

Yet this attempt to frame a Grand Universal Theory of Everything is itself not only highly ambitious but profoundly traditional. It flows from just the same kind of misplaced confidence which led physicists in the pre-modern, Aristotelian tradition to extend purposive reasoning beyond the sphere of human conduct, where it worked well, to explain the behaviour of stones, where it did not. Still more damagingly for Dennett's claims, it also resembles closely the vast metaphysical structures which Herbert Spencer built by extrapolating evolutionary ideas to all possible subject matters, thus producing, as his followers admiringly said, 'the theory of evolution dealing with the universe *as a whole*, from gas to genius'.[8] Darwin, though he remained polite in public, hated 'magnificent visions' of this kind. As he wrote in his *Autobiography*,

> I am not conscious of having profited in my work from Spencer's writings. His deductive manner of treating every subject is wholly opposed to my frame of mind. His conclusions never convince me . . . They partake more of the nature of definitions than of laws of nature.[9]

In short, Darwin understood that large ideas do indeed become dangerous if they are inflated beyond their proper use – dangerous to honesty, to intelligibility, to all the proper purposes of thought. For him the concept of natural selection was strictly and solely a biological one and even in biology he steadily rejected the claim that it was a universal explanation. He re-emphasised this point strongly in the sixth edition of the *Origin*:

> As my conclusions have lately been much misrepresented, and it has been stated that I attribute the modification of species exclusively to natural selection, I may be permitted to remark that in the first edition of this work, and subsequently, I placed in a most conspicuous position – namely at the close of the Introduction – the following words. 'I am convinced that natural selection has been the main, but not the exclusive, means of modification'. This has been of no avail. Great is the power of steady misrepresentation.[10]

The Search for Scientific Façades

Where does this history leave us today? How can we fit the science that is now so important to us into the general pattern of our lives without distorting anything? People find it hard now to imitate the caution of the founders of modern science, who managed to keep their science in isolation and carefully avoided applying physical concepts to any mental or social questions. But now that this science is so successful the temptation to expand its empire is much stronger. The social sciences have for some time tried to acquire its coinage in more or less realistic ways. That coinage, however, has gradually been redesigned in a way that disqualifies such realistic methods from counting as science. Standards are now set which concentrate on form, not on suitability to the subject-matter, so it becomes necessary to use methods which imitate scientific forms. And among those forms a prime favourite is, of course, atomism.

On this principle, then, if we want to understand culture we must somehow find its units. But is culture the sort of thing that divides up into units? Edward O. Wilson boldly declares that it is. In his book *Consilience*,[11] which seriously tries to mend the culture gap, he proposes this atomisation as the means of reconciling the humanities and social sciences with science by bringing them within its province. Culture (he says) must be atomisable because atomising is the way in which we naturally think:

> The descent to minutissima, the search for ultimate smallness in entities such as electrons, is a driving impulse of Western natural science. It is a kind of instinct. Human beings are obsessed with building blocks, for ever pulling them apart and putting them back together again . . . The impulse goes back as far as 400 BC . . . when Leucippus and Democritus speculated, correctly as it turned out, that matter is made of atoms. (p. 53)

400 BC scarcely seems long enough to certify an instinct. Granted, however, that people do often break things into units and that this is sometimes useful, is culture a suitable candidate for the treatment? Well, says Wilson, it has to be understood somehow. But what does it mean to understand it? Wilson reveals his odd stance here by saying that the best way to understand culture would not be to investigate the thoughts and intentions of the people practising it, but to know how it developed in the course of evolution. Since, however, that is impossible the next best way of understanding 'gene-culture co-evolution' must be

> to search for the basic unit of culture . . . Such a focus may seem at first contrived and artificial, but it has many worthy precedents. The great success of the natural sciences has been achieved substantially by the reduction of each physical phenomenon to its constituent elements followed by the use of the elements to reconstitute the holistic properties of the phenomenon. (p. 147)

Again, this argument reproduces, in a reverse direction, the same mistake which Aristotelian physics made when it extended explanation by purpose from the human sphere to the sphere of inanimate matter. Stones do not have purposes, but neither do cultures have particles. The example of physics cannot justify imposing its scheme on a quite different subject-matter.

Atoms or Genes?

Are there, perhaps, reasons of conceptual convenience forcing us to impose this apparently unsuitable pattern on thought? This must depend on what we are trying to do, and various memologists seem to have different aims. At times, Wilson himself clearly means to keep quite close to the pattern set by the discovery of physical particles. He hopes to find minutissimae, ultimate units of thought and to connect them eventually with particular minimal brain-states so as to provide (as he says) a kind of alphabet of a brain-language underlying all thought. This is an almost inconceivably ambitious project, a wild kind of cosmic expansion of Leibniz's quest for a universal language. But it is unmistakably a search for units of thought, not for units of culture. As he says, 'I have faith that the unstoppable neuroscientists will . . . in due course . . . capture the physical basis of mental concepts through the mapping of neural activity patterns.' (p. 148)

At other times, however, Wilson forgets this project entirely and describes his particles just as readily as 'units of culture'. And the examples that other memologists give mostly conform to this quite different model. Richard Dawkins, the original memologist, lists as his 'units of cultural transmission' 'tunes, ideas, catch-phrases, clothes-fashions, ways of making pots or of building arches', to which he adds popular songs, stiletto heels, the idea of God and Darwinism – certainly not the kind of things which could figure as Wilsonian ultimate units of thought.[12] Dawkins, however, insists that they are not merely convenient divisions of culture either but fixed, distinct natural units:

> There is something, some essence of Darwinism, which is present in the heads of every individual who understands the theory. If this were not so, then almost any statement about two people agreeing with each other would be meaningless. An 'idea-meme' might be defined as an entity which is capable of being transmitted from one brain to another . . . *The differences in the way that people represent the theory are then by definition, not part of the meme.* (p. 210, emphasis mine)

Memes are still intended, then, as *minutissima*, ultimate divisions, though here they are particles of culture rather than of thought. Daniel Dennett is equally insistent that these units are distinct and lasting natural divisions not just conventional ones. 'These new replicators are, roughly, ideas . . . the sort of complex ideas that form themselves into distinct memorable units.'[13] Giving a list of examples, even more mixed than Dawkins's, in which he includes deconstructionism, the *Odyssey* and wearing clothes, Dennett comments:

> Intuitively we see these as more or less identifiable cultural units, but we can say something more precise about how we draw the boundaries . . . the units are *the smallest elements that replicate themselves with reliability and fecundity*. We can compare them, in this regard, to genes and their components . . . A three-nucleotide phrase does not count as a gene for the same reason that you can't copyright a three-note musical phrase. (p. 344, emphasis mine)

But the law of copyright, being an artefact devised for civic convenience, is not much use in an attempt to establish fixed natural units. Wearing clothes is not any sort of minimum unit but a general term used to cover a vast range of customs. Deconstructionism is a loose name covering a group of ideas which stand in some sort of historical relation, a group with no fixed core. Darwinism only looks more plausible because of its unifying reference to Darwin. It too is a very complex group of ideas with no certified minimum core. Notoriously, Marx said that he was not a Marxist and (as we have seen) Darwin might well want to take the same line were he alive today. Views about what is central to such groupings vary widely and are normative, not factual. Doctrines that are at all interesting are wide patterns of thinking, not atomic beliefs. As dictionary makers find, they usually cannot be defined by any single nugget of meaning. Again, the *Odyssey* contains many elements memorable on their own, such as the stories of the Cyclops, of Scylla and Charybdis, and of the Wandering Rocks, so it can hardly be a minimal unit.

What is now the point of the proposal? If memes really correspond to Dawkinsian genes they must indeed be fixed units – hidden, unchanging causes of the changing items that appear round us in the world. But all the examples we are given correspond to phenotypes. They are the apparent items themselves. Moreover, most of the concepts mentioned cannot possibly be treated as unchanging or even as moderately solid. Such customs and ways of thinking are organic parts of human life, constantly growing, developing, changing and sometimes decaying like every other living thing. Much of this change, too, is due to our own action, to our deliberately working to change them. In one of his characteristic sudden spasms of acute critical insight Dennett himself notes this difficulty:

> Minds (or brains) . . . aren't much like photocopying machines at all. On the contrary, instead of just passing on their messages, correcting most of the typos as they go, brains seem to be designed to do just the opposite, to transform, invent, interpolate, censor, and generally mix up the 'input' before yielding any 'output' . . . . We *seldom* pass on a meme unaltered . . . Moreover, as Steven Pinker has stressed . . . much of the mutation that happens to memes (how much is not clear) is manifestly *directed* mutation. 'Memes such as the theory of relativity are not the cumulative product of millions of *random* (undirected) mutations of some original idea, but each brain in the chain of production added huge dollops of value to the product in a nonrandom way.' . . . Moreover, when memes come into contact with each other in a mind, they have a marvellous capacity to become adjusted to each other, swiftly changing their phenotypic effects to fit the circumstances. (pp. 354–5, author's emphasis)

So what, if anything, does this leave of the original parallel with genetics? How seriously is that parallel now intended? If memes are indeed something parallel to genes, as the last sentence of this quotation certainly implies – if they are hidden causes of culture rather than its units – what sort of entities are these causes supposed to be? They are not physical objects. But neither are they thoughts of the kind that normally play any part in our experience. They seem to be occult causes of those thoughts. How then do they manifest themselves? How do we know that they are there? It does not help to say that they are bits of information located in the infosphere (p. 347). Information is not a third kind of stuff, not an extra substance added to Cartesian mind and body or designed to supersede them. It is an abstraction from them. Invoking such an extra stuff is as idle as any earlier talk of phlogiston or animal spirits or occult forces. Information is facts about the world and we need to know where, in that world, these new and causally effective entities are to be found.

A Salutary Way of Thinking?

Without that knowledge, the parallel between memes and genes surely vanishes and the claim to scientific status with it. Meme-language is not really an extension of physical science but, as so often happens, an analogy which is welcomed, not for scientific merit but for moral reasons, as being a salutary way of thinking. At one point Dawkins himself speaks of it simply as an analogy 'which I find inspiring but which can be taken too far if we are not careful'.[14] Dennett, while making much stronger claims to scientific status, also adds that 'whether or not the meme perspective can be turned into science, in its philosophical guise it has already done much more good than harm' (p. 368). What kind of good? Dennett explains that the idea of memes corrects our tendency to exaggerate our own powers. It reminds us that we are not, as we 'would like to think, godlike creators of ideas, manipulating them and controlling them as our whim dictates and judging them from an independent, Olympian standpoint'. As he rightly says, we are not always 'in charge' (p. 346).

The relapsed parapsychologist Susan Blackmore, who has lately taken up the cause of memes, gives this moral point a special twist by grafting it, somewhat unexpectedly, on to the Buddhist doctrine that the self is an illusion: 'We all live our lives as a lie . . . belief in a permanent self is the cause of all human suffering.'[15] But such dismissals of everyday concepts have a quite different meaning according to what new item the dismisser offers as a substitute. What Buddhism offers is a deeper freedom, one which is held to flow from abandoning stereotypes about oneself and recognising the 'Buddha nature' within one. This nature is held to unite all living beings, without compromising their individual power to feel and act. It thus calls on us strongly to live in harmony with the rest of creation. By contrast, memetics offers only the news that we are (as Blackmore herself puts it) 'meme machines', constructions produced by alien viruses for their own purposes and incapable of having any purposes of their own. If anyone actually did try to believe this it is hard to see what could follow other than helpless fatalism, quickly followed by general breakdown. It is clear that the suggestion is, like so many other learned suggestions about selves, merely a paper doctrine about other people, not one by which anyone could live.

The chief reason why Blackmore accepts a belief in memes seems to be that she thinks it is the only possible alternative to Descartes's crude idea of a substantial self co-extensive with consciousness. To replace it she commends Dennett's anti-Cartesian model of the self, as proposed in his book *Consciousness Explained*. That model is a genuine attempt to depict the mind's own creative activity, but it cannot serve Blackmore's purpose. It leaves no room for memes and cannot accommodate them, even though Dennett himself has since taken up meme-talk. Dennett's remark that the idea of the self is a 'benign user illusion' misleads her. It is actually only a bit of residual Cartesian dualism, a suggestion by

Dennett that the 'self' is always conceived as a Cartesian disembodied ghost. 'Self' is in fact a highly complex idea with many different uses. Nothing can be gained, morally or metaphysically, by trying to shoot it down in favour of this tinpot successor.

In general, the moral point which she shares with Dennett – the demand for a correction of human vanity, the insistence on a more realistic notion of our species' place in nature – is a healthy and reasonable one. It is quite true that Western culture has systematically exaggerated both the power and the importance of *Homo sapiens* relative to the rest of creation. Thinkers like Dawkins and Wilson have done really useful work in correcting this absurdity, in making us more aware of our relative insignificance both in time, in the huge evolutionary perspective, and in the vast array of life forms that still surrounds us.

The value of that correction, however, depends on the reality of the particular causal background which is then introduced to replace human activity. About evolution, the correction works because it points to real forces in the world which are responsible for the results that people had supposed were due to human effort. In order to make the meme proposal parallel to this case, it would be necessary to show that memes, too, were genuine external forces, alien puppet-masters previously hidden from us but revealed now as the true causes ruling our life. That indeed is the dramatic picture which Dawkins originally suggested, quoting a remark of Nick Humphrey: 'Memes should be regarded as living structures, *not just metaphorically but technically*. When you plant a fertile meme in my mind you *literally* parasitise my brain, turning it into a vehicle for the meme's propagation in just the way that a virus may parasitise the genetic mechanism of a host cell.'[16] (Emphases mine.)

This was the sort of language that made the proposal seem so exciting and important in the first place. If the alleged discovery had been a real one, it would indeed have been important – but also, of course, disastrous since it would have entailed total fatalism. Dennett tries to disinfect the imagery somewhat by shifting the metaphor to symbiosis, citing as a close parallel, 'the creation of eukaryotic cells that made multicellular life possible . . . one day some prokaryotes were invaded by parasites of sorts and this turned out to be a blessing in disguise, for . . . these invaders turned out to be beneficial and hence were *symbionts* but not parasites.' (p. 340)

But however soothing this change may be emotionally it still does not give these entities any sort of intelligible status. In order to conceive ideas, or their mental causes, as separate organisms existing in their own right before infesting minds, we would need to forsake empiricism and build a very bold – perhaps Hegelian? – framework of objective idealism, allowing mental entities this independent status outside particular minds. And idealism is as far as possible from Dennett's philosophical style. It seems extraordinary that a thinker as committed as he is to the continuity of evolution should choose to build this rickety metaphysical edifice in order to keep mental entities separate from us instead of treating our thoughts and customs as what they obviously are – namely, forms of activity which our species has gradually developed during its history to supply its needs. As William of Occam observed, varieties of entities should not be multiplied beyond necessity. When human beings think and act, no entities need to be present in them except those thinking and acting human beings themselves.

Why does Dennett insist in this way on the memes' independent existence? He explains that his main point is that our thoughts do not always do us any good and must therefore not be thought of as (he rather surprisingly says) humanists think of them – as entities aiming at our advantage, but as aiming at their own:

> The meme's eye perspective challenges one of the central axioms of the humanities . . . we tend to overlook the fundamental fact that 'a cultural trait may have evolved in the way it has simply because it is *advantageous to itself*'. (p. 362, author's emphasis)

> Competition is the major selective force in the infosphere and, just as in the biosphere, the challenge has been met with great ingenuity . . . Like a mindless virus, a meme's prospects depend on its design – not its internal design, whatever that might be, but the design it shows the world, its phenotype, the way it affects things in its environment [namely] minds and other memes. (p. 349)

Memetics, then, is needed to help us understand the strategies by which memes contrive to infest us even when they are not useful to us, for example, 'the meme for faith, which discourages the exercise of the sort of critical judgement that might decide that the idea of faith was, all things considered, a dangerous idea'. (p. 349)

Thus (it seems) if we want to know why people have faith in something – for instance, why Western people today often have faith in the pronouncements of scientists – we ought not to ask what reasons, good or bad, these people have for that confidence. Instead, we should simply note that the idea of faith is an efficient parasite. But how would that get us any further?

Motivation Is Not a New Topic

At this point we need to say something obvious. The fact that our thoughts and customs are not always to our advantage is not a new scientific discovery. It is a most familiar platitude, both in everyday life and in traditional humanistic thinking. We know all too well that our thoughts and customs often lead us to act foolishly, destructively, even suicidally. And the crucial point about this self-destructive tendency – the thing that makes it peculiarly distressing – is that in these cases the conflicting motives which lead to the trouble are indeed all our own. They do *not* arise from possession by some kind of external parasite. They are warring parts of ourselves.

Far from this recognition being alien to the humanities, it has always been one of their central themes. It is central in literature, where it lies at the root of both tragedy and comedy. And in history our interest in this human tendency to self-destructiveness is crucial because it directs our curiosity about the past – because we need to know why things go so badly wrong much more urgently than we need to celebrate our successes. It is also the starting point of our reflection about the deep practical dilemmas that give rise to moral philosophy.

In the humanistic disciplines so far, enquiry about self-destructive behaviour has mostly concentrated on the attempt to understand human motivation. That is not at all the area to which memologists direct our attention, but it is one that does indeed hold hidden causes of thought and action. Those causes, however, are not hidden in the sense in which DNA was hidden before it was discovered. They are not facts in the outer world which merely happen not to have been researched yet: they are facts about motives, and they are obscure largely because we find it so hard and painful to attend to them. They are facts which we cannot understand properly unless we are prepared to make some serious imaginative effort of identification with the actors in question.

That is why literature is such an important part of our lives – why the notion that it is less important than science is so mistaken. Shakespeare and Tolstoy help us to understand the self-destructive psychology of despotism. Flaubert and Racine illuminate the self-destructive side of love. What we need to grasp in such cases is not the simple fact that people are acting against their interests. We know that; it stands out a mile. We need to understand, beyond this, what kind of gratification they are getting from acting in this way. If, instead of looking for this factor directly and imaginatively by studying their conduct, we were to shift our attention to the alleged interactions between populations of memes, as Dennett advises, we would lose a crucially important source of knowledge in order to pursue a phantom.

Explaining Witch-Hunting

The effect of this exchange can be interestingly seen by looking at the kind of example where, at a casual glance, we might find it most persuasive, namely in cases where large numbers of people do act irrationally for motives which seem really obscure. For instance, consider the witch craze that prevailed in Europe from the fifteenth to the seventeenth centuries. This craze was not, as is often supposed, simply a survival of ancient superstition caused by ignorance and finally cured by the rise of science. On the contrary, in the Middle Ages there were few prosecutions for witchcraft because the Church authorities thought that witchcraft was rare (though real) and they discouraged witch-hunting because they saw the danger of false accusation. It was in the Renaissance that things changed. At that time, as a recent historian puts it:

> The Europeans did three things which set them far apart from most other peoples at most other times and places. Between 1500 and 1700 they set sail in tall ships and colonised the far corners of the globe. They made stunning strides forward in the sciences. And they executed tens of thousands of people, mainly women, as witches.[17]

The attack of frenzy coincided, then, with the increase of knowledge rather than being cured by it. And, as these authors show, when it finally subsided it did not do so because science had shown that bewitchment was physically impossible, but because people gradually came to find it psychologically incredible that there was such an organised host of demon worshippers. Writers of various kinds greatly helped to nourish this incredulity, but scientific arguments do not seem to have contributed anything particular to it.

I cite this case because (as I say) it is one which really does need explanation and one where explanation by memes would look so easy. We need only posit a new meme successfully invading a population that has no immunity to it, a meme which declines later as that immunity develops. Its success is then due to its own reproductive strategy – presumably produced by a mutation – not to any fact about the people concerned. We need not look at those people. We need not relate the meme to these people's intentions. We certainly need not look at human psychology generally and look into our own hearts to see what we might learn there about such conduct. We simply place the whole causation outside human choice, thus avoiding that overestimation of our own powers which so disturbs Dennett. This effort to avoid pride would of course land us in a quite unworkable kind of fatalism.

But, on top of that, the meme story simply fails to give us any kind of explanation at all. What we need to understand in such a case is how people could begin to think and act in this way in spite of the beliefs, customs and ideals which had prevented them from doing so earlier. We need, in fact, to understand the psychology of persecution and xenophobia. We need to penetrate paranoia. We need this not just in relation to the witch craze but for understanding the oddities of human conduct at other times and places too, not least in our own lives. Understanding it does not mean discovering, by research, new facts about the behaviour of an imaginary alien life-form. It means essentially self-knowledge, an exploration of what de Tocqueville called 'the habits of the heart'. Examining the evolutionary strategies of mythical culture units cannot save us from this awkward form of investigation.

Notes and References

1 Richard Dawkins, *The Selfish Gene* (Oxford, Oxford University Press, 1976), ch. 11, pp. 203–15.

2 John Lyons (ed.), *New Horizons In Linguistics* (Harmondsworth, Penguin, 1970). Editor's abstract introducing Chapter 4 on 'Phonology' by E. C. Fudge, p. 76.

3 See a good discussion of tropisms in Robert Boakes, *From Darwin to Behaviourism* (Cambridge, Cambridge University Press, 1984), pp.138–40.

4 C. B. A. Behrens, *The Ancien Regime* (London, Thames & Hudson, 1967), pp. 123–4.

5 David Hume, *Enquiry Concerning the Principles of Morals* (1751), paragraphs 163 and 192.

6 Especially in *Darwin's Dangerous Idea* (Harmondsworth, Penguin, 1996). Page references henceforth refer to this book, though Dennett has of course discussed the matter in other places.

7 See my article 'One World But a Big One', *Journal of Consciousness Studies*, vol. 3, nos. 5–6 (1996), pp. 500–15.

8 Edward Clodd, Spencer's follower and interpreter, thus triumphantly described his achievement. See A. C. Armstrong, *Transitional Eras in Thought, with Special Reference to the Present Age* (New York, Macmillan, 1904), p. 48.

9 *The Autobiography of Charles Darwin, 1809–1882, with Original Omissions Restored*, ed. Nora Barlow (New York, Harcourt, Brace, 1958), p. 109.

10 Sixth edition (1872), p. 395.
11 E. O. Wilson, *Consilience: The Unity of Knowledge* (London, Little Brown, 1998).
12 Dawkins, *The Selfish Gene*, pp. 207–9.
13 Dennett, *Darwin's Dangerous Idea*, p. 344.
14 Richard Dawkins, *The Blind Watchmaker* (Harlow, Longman, 1986), p. 196.
15 Susan Blackmore, 'Meme, Myself, I', *New Scientist*, 2177 (13 March 1999), pp. 40–4. See also her book *The Meme Machine* (Oxford, Oxford University Press, 1999).
16 Dawkins, *The Selfish Gene*, p. 207.
17 Karen Green and John Bigelow, 'Does Science Persecute Women? The Case of the 16th–17th Century Witch-Hunts', *Philosophy*, vol. 73, no. 284 (April 1998), p. 199.

6

More Things in Heaven and Earth

Stephen Jay Gould

Darwinian Fundamentalism

With copious evidence ranging from Plato's haughtiness to Beethoven's tirades, we may conclude that most brilliant people of history tend to be a prickly lot. But Charles Darwin must have been the most genial of geniuses. He was kind to a fault, even to the undeserving, and he never uttered a harsh word – or hardly ever, as his countryman Captain Corcoran once said. Darwin's disciple, George Romanes, expressed surprise at the only sharply critical Darwinian statement he had even encountered: 'In the whole range of Darwin's writings there cannot be found a passage so strongly worded as this: it presents the only note of bitterness in all the thousands of pages which he has published.' Darwin directed the passage which Romanes found so striking against people who would simplify and caricature his theory as claiming that natural selection, and only natural selection, caused all evolutionary changes. He wrote in the last (1872) edition of *The Origin of Species*:

> As my conclusions have lately been much misrepresented, and it has been stated that I attribute the modification of species exclusively to natural selection, I may be permitted to remark that in the first edition of this work and subsequently, I place in a most conspicuous position – namely at the close of the Introduction – the following words: 'I am convinced that natural selection has been the main but not the exclusive means of modification.' This has been of no avail. Great is the power of steady misrepresentation.

Darwin clearly loved his distinctive theory of natural selection – the powerful ideas that he often identified in letters as his dear 'child'. But, like any good parent, he understood limits and imposed discipline. He

85

knew that the complex and comprehensive phenomena of evolution could not be fully rendered by any single cause, even one so ubiquitous and powerful as his own brainchild.

In this light, especially given history's tendency to recycle great issues, I am amused by an irony that has recently ensnared evolutionary theory. A movement of strict constructionism, a self-styled form of Darwinian fundamentalism, has risen to some prominence in a variety of fields, from the English biological heartland of John Maynard Smith to the uncompromising ideology (albeit in graceful prose) of his compatriot Richard Dawkins, to the equally narrow and more ponderous writing of the American philosopher Daniel Dennett.[1] Moreover, a larger group of strict constructionists are now engaged in an almost mordantly self-conscious effort to 'revolutionise' the study of human behaviour along a Darwinian straight and narrow under the name of 'evolutionary psychology'.

Some of these ideas have filtered into the general press, but the uniting theme of Darwinian fundamentalism has not been adequately stressed or identified. Professionals, on the other hand, are well aware of the connections. My colleague Niles Eldredge, for example, speaks of this co-ordinated movement as ultra-Darwinism.[2] Amid the variety of their subject matter, the ultra-Darwinists share a conviction that natural selection regulates everything of any importance in evolution, and that adaptation emerges as a universal result and ultimate test of selection's ubiquity.

The irony of this situation is twofold. First, as illustrated by the quotation above, Darwin himself strongly opposed the ultras of his own day. (In one sense, this nicety of history should not be relevant to modern concerns: maybe Darwin was overcautious, and modern ultras therefore out-Darwin Darwin for good reason. But since the modern ultras push their line with an almost theological fervour, and since the views of founding fathers do matter in religion, though supposedly not in science, Darwin's own fierce opposition does become a factor in judgement.) Second, the invigoration of modern evolutionary biology with exciting nonselectionist and nonadaptationist data from the three central disciplines of population genetics, developmental biology and palaeontology (see examples below) makes our pre-millennial decade an especially unpropitious time for Darwinian fundamentalism – and seems only to reconfirm Darwin's own eminently sensible pluralism.

Charles Darwin often remarked that his revolutionary work had two distinct aims: first, to demonstrate the fact of evolution (the genealogical connection of all organisms and a history of life regulated by 'descent with modification'); second, to advance the theory of natural selection as the most important mechanism of evolution. Darwin triumphed in his first aim (American creationism of the Christian far right notwithstanding). Virtually all thinking people accept the factuality of evolution and no conclusion in science enjoys better documentation. Darwin also succeeded substantially in his second aim. Natural selection, an immensely powerful idea with radical philosophical implications, is surely a major cause of evolution as validated in theory and demonstrated by countless experiments. But is natural selection as ubiquitous and effectively exclusive as the ultras propose?

The radicalism of natural selection lies in its power to dethrone some of the deepest and most traditional comforts of Western thought, particularly the notion that nature's benevolence, order, and good design, with humans at a sensible summit of power and excellence, prove the existence of an omnipotent and benevolent creator who loves us most of all (the old-style theological version), or at least that nature has meaningful directions and that humans fit into a sensible and predictable pattern regulating the totality (the modern and more secular version).

To these beliefs Darwinian natural selection presents the most contrary position imaginable. Only one causal force produces evolutionary change in Darwin's world: the unconscious struggle among individual organisms to promote their own personal reproductive success – nothing else, and nothing higher (no force, for example, works explicitly for the good of species or the harmony of ecosystems). Richard Dawkins would narrow the focus of explanation even one step further – to genes struggling for reproductive success within passive bodies (organisms) under the control of genes – a hyper-Darwinian idea that I regard as a logically flawed and basically foolish caricature of Darwin's genuinely radical intent.

The very phenomena that traditional views cite as proof of benevolence and intentional order – the good design or organisms and the harmony of ecosystems – arise by Darwin's process of natural selection only as side consequences of a singular causal principle of apparently opposite meaning: organisms struggling for themselves alone. (Good design becomes one pathway to reproductive success, while the harmony of ecosystems records a competitive balance among victors.) Darwin's system should be viewed as morally liberating, not cosmically depressing. The answers to moral questions cannot be found in nature's factuality in any case, so why not take the 'cold bath' of recognising nature as nonmoral, and not constructed to match our hopes? After all, life existed on earth for 3.5 billion years before we arrived; why should life's causal ways match our prescriptions for human meaning or decency?

We now reach the technical and practical point that sets the ultra-Darwinian research agenda. Natural selection can be observed directly, but only in the unusual circumstances of controlled experiments in laboratories (on organisms with very short generations such as fruit flies) or within simplified and closely monitored systems in nature. Since evolution, in any substantial sense, takes so much time (more than the entire potential history of human observing!), we cannot, except in special circumstances, watch the process in action, and must therefore try to infer causes from results – the standard procedure in any historical science, by the way, and not a special impediment facing evolutionists.

The generally accepted result of natural selection is adaptation – the shaping of an organism's form, function and behaviour to achieve the Darwinian *summum bonum* of enhanced reproductive success. We must therefore study natural selection primarily from its results – that is, by concentrating on the putative adaptations of organisms. If we can interpret all relevant attributes of organisms as adaptations for reproductive success, then we may infer that natural selection has been the cause of evolutionary change. This strategy of research – the so-called adaptationist programme – is the heart of Darwinian biology and the fervent, singular credo of the ultras.

Since the ultras are fundamentalists at heart, and since fundamentalists generally try to stigmatise their opponents by depicting them as apostates from the one true way, may I state for the record that I (along with all other Darwinian pluralists) do not deny either the existence and central importance of adaptation, or the production of adaptation by natural selection. Yes, eyes are for seeing and feet are for moving. And, yes again, I know of no scientific mechanism other than natural selection with the proven power to build structures of such eminently workable design.

But does all the rest of evolution – all the phenomena of organic diversity, embryological architecture and genetic structure, for example – flow by simple extrapolation from selection's power to create the good design of organisms? Does the force that makes a functional eye also explain why the world houses more than 500,000 species of beetles and fewer than fifty species of priapulid worms? Or why most nucleotides in multicellular creatures do not code for any enzyme or protein involved in the construction of an organism? Or why ruling dinosaurs died and subordinate mammals survived to flourish and, along one oddly contingent pathway, to evolve a creature capable of building cities and understanding natural selection?

I do not deny that natural selection has helped us to explain phenomena at scales very distant from individual organisms, from the behaviour of an ant colony to the survival of a redwood forest. But selection cannot suffice as a full explanation for many aspects of evolution: for other types and styles of causes become relevant, or even prevalent, in domains both far above and far below the traditional Darwinian locus of the organism. These other causes are not, as the ultras often claim, the product of thinly veiled attempts to smuggle purpose back into biology. These additional principles are as direction-less, non-teleological and materialistic as natural selection itself – but they operate differently from Darwin's central mechanism. In other words, I agree with Darwin that natural selection is 'not the exclusive means of modification'.

What an odd time to be a fundamentalist about adaptation and natural selection – when each major subdiscipline of evolutionary biology has been discovering other mechanisms as adjuncts to selection's centrality. Population genetics has worked out in theory, and validated in practice, an elegant, mathematical account of the large role that neutral, and therefore nonadaptive, changes play in the evolution of nucleotides, or individual units of DNA programmes. Eyes may be adaptations, but most substitutions of one nucleotide for another within populations may not be adaptive.

In the most stunning evolutionary discoveries of the past decade, developmental biologists have documented an astonishing 'conserva-tion', or close similarity, of basic pathways of development among phyla that have been evolving independently for at least 500 million years, and that seem so different in basic anatomy (insects and vertebrates, for example). The famous homeotic genes of fruit flies – responsible for odd mutations that disturb the order of parts along the main body axis, placing legs, for example, where antennae or mouth parts should be – are also present (and repeated four times on four separate chromosomes) in vertebrates, where they function in effectively the same way. The major developmental pathway for eyes is conserved and mediated by the same gene in squids, flies and vertebrates, though the end products differ substantially (our single-lens eye versus the multiple facets of insects). The same genes regulate the formation of top and bottom surfaces in insects and vertebrates, though with inverted order – as our back, with the spinal cord running above the gut, is anatomically equivalent to an insect's belly, where the main nerve cords run along the bottom surface, with the gut above.

One could argue, I suppose, that these instances of conservation only record adaptation, unchanged through all of life's vicissitudes because their optimality can't be improved. But most biologists feel that such stability acts primarily as a constraint upon the range and potentiality of adaptation, for if organisms of such different function and ecology must build bodies along the same basic pathways, then limitation of possibilities rather than adaptive honing to perfection becomes a dominant theme in evolution. At a minimum, in explaining evolutionary pathways through time, the constraints imposed by history rise to equal prominence with the immediate advantages of adaptation.

My own field of palaeontology has strongly challenged the Darwinian premise that life's major transformations can be explained by adding up, through the immensity of geological time, the successive tiny changes produced generation after generation by natural selection. The extended stability of most species, and the branching off of new species in geological moments (however slow by the irrelevant scale of a human life) – the pattern known as punctuated equilibrium – require that long-term evolutionary trends be explained as the distinctive success of some species versus others, and not as a gradual accumulation of adaptations generated by organisms within a continuously evolving population. A trend may be set by high rates of branching in certain species within a larger group. But individual organisms do not branch; only populations do – and the causes of a population's branching can rarely be reduced to the adaptive improvement of its individual organisms.

The study of mass extinction has also disturbed the ultra-Darwinian consensus. We now know, at least for the terminal Cretaceous event some 65 million years ago which wiped out dinosaurs along with about 50 per cent of marine invertebrate species, that some episodes of mass extinction are both truly catastrophic and triggered by extraterrestrial impact. The death of some groups (like dinosaurs) in mass extinctions and the survival of others (like mammals), while surely not random, probably bears little relationship to the evolved, adaptive reasons for success of lineages in normal Darwinian times dominated by competition. Perhaps mammals survived (and humans ultimately evolved) because small creatures are more resistant to catastrophic extinction. And perhaps Cretaceous mammals were small primarily because they could not compete successfully in the larger size ranges of dominant dinosaurs. Immediate adaptation may bear no relationship to success over immensely long periods of geological change.

Why then should Darwinian fundamentalism be expressing itself so stridently when most evolutionary biologists have become more pluralistic in the light of these new discoveries and theories? I am no psychologist, but I suppose that the devotees of any superficially attractive cult must dig in when a general threat arises. 'That old-time religion; it's good enough for me.' There is something immensely beguiling about strict adaptationism – the dream of an underpinning simplicity for an enormously complex and various world. If evolution were powered by a single force producing one kind of result, and if life's long messy history could therefore be explained by extending small or orderly increments of adaptation through the immensity of geological time, then an explanatory simplicity might descend upon evolution's overt richness. Evolution then might become 'algorithmic', a surefire logical procedure, as in Daniel Dennett's reverie. But what is wrong with messy richness, so long as we can construct an equally rich texture of satisfying explanation?

Daniel Dennett's 1995 book, *Darwin's Dangerous Idea*, presents itself as the ultras' philosophical manifesto of pure adaptationism. Dennett explains the strict adaptationist view well enough, but he defends a blinkered picture of evolution in assuming that all important phenomena can be explained thereby.

Dennett bases his argument on three images or metaphors, all sharing the common error of assuming that conventional natural selection, working in the adaptationist mode, can account for all evolution by extension – so that the entire history of life becomes one grand solution to problems in design. 'Biology is engineering,' Dennett tells us again and again. In a devastating review, published in the leading professional journal *Evolution*, and titled 'Dennett's Dangerous Idea', H. Allen Orr notes:

> His review of attempts by biologists to circumscribe the role of natural selection borders on a zealous defense of panselectionism. It is also absurdly unfair . . . Dennett fundamentally misunderstands biologists' worries about adaptationism. Evolutionists are essentially unanimous that – where there is 'intelligent Design' – it is caused by natural selection . . . Our problem is that, in many adaptive stories, the protagonist does not show dead-obvious signs of Design.

In his first metaphor Dennett describes Darwin's dangerous idea of natural selection as a 'universal acid' – to honour both its ubiquity and its power to corrode traditional Western beliefs. Speaking of adaptation, natural selection's main consequence, Dennett writes, 'It plays a crucial role in the analysis of every biological event at every scale from the creation of the first self-replicating macromolecule on up.' I certainly accept the acidic designation – for the power and influence of the idea of natural selection does lie in its radical philosophical content – but few biologists would defend the blithe claim for ubiquity. If Dennett chooses to restrict his personal interest to the engineering side of biology – the part that natural selection does construct – then he is welcome to do so. But he may not impose this limitation upon others, who know that the record of life contains many more evolutionary things than are dreamt of in Dennett's philosophy.

Natural selection does not explain why many evolutionary transitions from one nucleotide to another are neutral, and therefore nonadaptive. Natural selection does not explain why a meteor crashed into the earth 65 million years ago, setting in motion the extinction of half the world's species. As Orr points out, Dennett's disabling parochialism lies most clearly exposed in his failure to discuss the neutral theory of molecular evolution, or even to mention the name of its founder, the great Japanese geneticist Motoo Kimura – for few evolutionary biologists would deny that this theory ranks among the most interesting and powerful adjuncts to evolutionary explanation since Darwin's formulation of natural selection. You don't have to like the idea, but how can you possibly leave it out?

In a second metaphor Dennett continually invokes an image of cranes and skyhooks. In his reductionist account of evolution cranes build the good design of organisms upward from nature's physicochemical substrate. Cranes are good. Natural selection is evolution's basic crane; all other cranes (sexual reproduction, for example) act as mere auxiliaries to boost the speed or power of natural selection in constructing organisms of good design. Skyhooks, on the other hand, are spurious forms of special pleading that reach down from the numinous heavens and try to build organic complexity with *ad hoc* fallacies and speculations unlinked to other proven causes. Skyhooks, of course, are bad. Everything that isn't natural selection, or an aid to the operation of natural selection, is a skyhook.

But a third (and correct) option exists to Dennett's oddly dichotomous Hobson's choice: either accept the idea of one basic crane with auxiliaries, or believe in skyhooks. May I suggest that the platform of evolutionary explanation houses an assortment of basic cranes, all helping to build the edifice of life's history in its full grandeur (not only the architecture of well-engineered organisms). Natural selection may be the biggest crane with the largest set of auxiliaries, but Kimura's theory of neutralism is also a crane; so is punctuated equilibrium; so is the channelling of evolutionary change by developmental constraints. 'In my father's house are many mansions' – and you need a lot of cranes to build something so splendid and variegated.

For his third metaphor – though he would demur and falsely label the claim as a fundamental statement about causes – Dennett describes evolution as an 'algorithmic process'. Algorithms are abstract rules of calculation, fully general in making no reference to particular content. In Dennett's words, 'an algorithm is a certain sort of formal process that can be counted on – logically – to yield a certain sort of result whenever it is "run" or instantiated'. If evolution truly works by an algorithm, then all else in Dennett's simplistic system follows: we need only one kind of crane to supply the universal acid.

I am perfectly happy to allow – indeed I do not see how anyone could deny – that natural selection, operating by its bare-bones mechanics, is algorithmic: variation proposes and selection disposes. So if natural selection builds all of evolution without the interposition of auxiliary processes or intermediary complexities, then I suppose that evolution is algorithmic too. But – and here we encounter Dennett's disabling error once again – evolution includes so much in addition to natural selection that it cannot be algorithmic in Dennett's simple calculational sense.

Yet Dennett yearns to subsume all the phenomenology of nature under the limited aegis of adaptation as an algorithmic result of natural selection. He writes, 'Here, then, is Darwin's dangerous idea: the algorithmic level *is* the level that best accounts for the speed of the antelope, the wing of the eagle, the shape of the orchid, the diversity of species, and all the other occasions for wonder in the world of nature' (Dennett's italics). I will grant the antelope's run, the eagle's wing, and much of the orchid's shape – for these are adaptations, produced by natural selection, and therefore legitimately in the algorithmic domain. But can Dennett really believe his own imperialistic extensions? Is the diversity of species no more than a calculational consequence of natural selection? Can anyone really believe, beyond the hype of rhetoric, that '*all* the other occasions for wonder in the world of nature' flow from adaptation?

Perhaps Dennett only gets excited when he can observe adaptive design, the legitimate algorithmic domain; but such an attitude surely represents an impoverished view of nature's potential interest. I regard the neutral substitution of nucleotides as an 'occasion for wonder in the world of nature'. And I marvel at the probability that the impact of a meteor wiped out dinosaurs and gave mammals a chance. If this contingent event had not occurred, and imparted a distinctive pattern to the evolution of life, we would not be here to wonder about anything at all!

The Fallacies of 'Evolutionary Psychology' and the Pleasures of Pluralism

Darwin began the last paragraph of *The Origin of Species* (1859) with a famous metaphor about life's diversity and ecological complexity:

> It is interesting to contemplate an entangled bank, clothed with many plants of many kinds, with birds singing on the bushes, with various insects flitting about, and with worms crawling through the damp earth, and to reflect that these elaborately constructed forms, so different from each other, and dependent on each other in so complex a manner, have all been produced by laws acting around us.

He then begins the final sentence of the book with an equally famous statement: 'There is grandeur in this view of life . . .'

For Darwin, as for any scientist, a kind of ultimate satisfaction (Darwin's 'grandeur') must reside in the prospect that so much variety and complexity might be generated from natural regularities – the 'laws acting around us accessible to our intellect and empirical probing'. But what is the proper relationship between underlying laws and explicit results? The fundamentalists among evolutionary theorists revel in the belief that one overarching law – Darwin's central principle of natural selection – can render the full complexity of outcomes (by working in conjunction with auxiliary principles, like sexual reproduction, that enhance its rate and power).

The 'pluralists', on the other hand – a long line of thinkers including Darwin himself, however ironic this may seem since the fundamentalists use the cloak of his name for their distortion of his position – accept natural selection as a paramount principle (truly *primus inter pares*), but then argue that a set of additional laws, as well as a large role for history's unpredictable contingencies, must also be invoked to explain the basic patterns and regularities of the evolutionary pathways of life. Both sides locate the 'grandeur' of 'this view of life' in the explanation of complex and particular outcomes by general principles, but ultra-Darwinian fundamentalists pursue one true way, while pluralists seek to identify a set of interacting explanatory modes, all fully intelligible, although not reducible to a single grand principle like natural selection.

Daniel Dennett devotes the longest chapter in *Darwin's Dangerous Idea* to an excoriating caricature of my ideas, all in order to bolster his defence of Darwinian fundamentalism. If an argued case can be discerned at all amid the slurs and sneers, it would have to be described as an effort to claim that I have, thanks to some literary skill, tried to

94

raise a few piddling, insignificant and basically conventional ideas to 'revolutionary' status, challenging what he takes to be the true Darwinian scripture. Dennett claims that I have promulgated three 'false alarms' as supposed revolutions against the version of Darwinism that he and his fellow defenders of evolutionary orthodoxy continue to espouse.

Dennett first attacks my view that punctuated equilibrium is the dominant pattern of evolutionary change in the history of living organisms. This theory, formulated by Niles Eldredge and me in 1972, proposes that the two most general observations made by palaeontologists form a genuine and primary pattern of evolution and do not arise as artefacts of an imperfect fossil record. The first observation notes that most new species originate in a geological 'moment'. The second holds that species generally do not change in any substantial or directional way during their geological lifetimes – usually a long period averaging five to ten million years for fossil invertebrate species. Punctuated equilibrium does not challenge accepted genetic ideas about the rates at which species emerge (for the geological 'moment' of a single rock layer may represent many thousand years of accumulation). But the theory does contravene conventional Darwinian expectations for gradual change over geological periods and does suggest a substantial revision of standard views about the causes of long-term evolutionary trends. For such trends must now be explained by the higher rates at which some species branch off from others, and the greater durations of some stable species as distinguished from others, and not as the slow and continuous transformation of entire populations.

In his second attack, Dennett denigrates the importance of nonadaptive side consequences ('spandrels' in my terminology) as sources for later and fruitful reuse. In principle, spandrels define the major category of important evolutionary features that do not arise as adaptations. Since organisms are complex and highly integrated entities, any adaptive change must automatically 'throw off' a series of structural byproducts – like the mould marks on an old bottle or, in the case of an architectural spandrel itself, the triangular space 'left over' between a rounded arch and the rectangular frame of wall and ceiling. Such byproducts may later be co-opted for useful purposes, but they didn't arise as adaptations. Reading and writing are now highly adaptive for humans, but the mental machinery for these crucial capacities must have originated as spandrels that were co-opted later, for the brain reached its current size and conformation tens of thousands of years before any human invented reading or writing.

Third, and finally, Dennett denies theoretical importance to the roles of contingency and chance in the history of life, a history that has few predictable particulars and no inherent directionality, especially given the persistence of bacteria as the most common and dominant form of life on Earth ever since their origin as the first fossilised creatures some 3.5 billion years ago.[3] Bacteria are biochemically more diverse, and live in a wider range of environments (including near-boiling waters, and pore spaces in rocks up to two miles beneath the earth's surface), than all other living things combined. The number of *E. coli* cells in the gut of each human being exceeds the total number of human beings that have ever lived.

These three concepts work as pluralistic correctives to both the poverty and limited explanatory power of the ultra-Darwinian research programme. Punctuated equilibrium requires that substantial evolutionary trends over geological time, the primary phenomenon of macroevolution, be explained by the greater long-term success of some species versus others within a group of species descended from a common ancestor. Such trends cannot be explained, as Darwinian fundamentalists would prefer, as the adaptive success of individual organisms in conventional competition extrapolated through geological time as the slow and steady transformation of populations by natural selection. The principle of spandrels, discussed at greater length later in this chapter, stresses the role that nonadaptive side consequences play in structuring the directions and potentials of future evolutionary change. Taken together, punctuated equilibrium and spandrels invoke the operation of several important principles in addition (and sometimes even opposed) to conventional natural selection working in the engineering mode that Dennett sees as the only valid mechanism of evolution.

My third pluralistic corrective to traditional theory does not invoke other principles in addition to natural selection, but rather stresses the limits faced by *any* set of general principles in our quest to explain the actual patterns of life's history. Crank your algorithm of natural selection to your heart's content, and you cannot grind out the contingent patterns built during the earth's geological history. You will get predictable pieces here and there (convergent evolution of wings in flying creatures), but you will also encounter too much randomness from a plethora of sources, too many additional principles from within biological theory and too many unpredictable impacts from environmental histories beyond biology (including those occasional meteors) – all showing that the theory of natural selection must work in concert with several other principles of change to explain the observed pattern of evolution.

The fallacy of Dennett's argument undermines his other imperialist hope that the universal acid of natural selection might reduce human cultural change to the Darwinian algorithm as well. Dennett, following Dawkins, tries to identify human thoughts and actions as 'memes', thus viewing them as units that are subject to a form of selection analogous to natural selection of genes. Cultural change, working by memetic selection, then becomes as algorithmic as biological change operating by natural selection on genes – thus uniting the evolution of organisms and thoughts under a single ultra-Darwinian rubric:

> According to Darwin's dangerous idea . . . not only all your children and your children's children, but all your brainchildren and your brain-children's brainchildren must grow from the common stock of Design elements, genes and memes . . . Life and all its glories are thus united under a single perspective.

But, as Dennett himself correctly and repeatedly emphasises, the generality of an algorithm depends upon 'substrate neutrality'. That is, the various materials (substrates) subject to the mechanism (natural selection in this case) must all permit the mechanism to work in the same effective manner. If one kind of substrate tweaks the mechanism to operate differently (or, even worse, not to work at all), then the algorithm fails. To choose a somewhat silly example which actually played an important role in recent American foreign policy, the Cold War 'domino theory' held that communism must be stopped every-where because if one country turned red, then others would do so as well, for countries are like dominoes standing on their ends and placed one behind the other – so that the toppling of one must propagate down the entire line to topple all. Now if you devised a general formula (an algorithm) to describe the necessary propagation of such toppling, and wanted to cite the algorithm as a general rule for all systems made of a series of separate objects, then the generality of your algorithm would depend upon substrate neutrality – that is, upon the algorithm's common working regardless of substrate (similarly for dominoes and nations in this case). The domino theory failed because differences in substrate affect the outcome, and such differences can even derail the operation of the algorithm. Dominoes must topple, but the second nation in a line might brace itself, stay upright upon impact, and therefore fail to propagate the collapse.

Natural selection does not enjoy this necessary substrate neutrality. As the great evolutionist R. A. Fisher showed many years ago in the founding document of modern Darwinism *(The Genetical Theory of*

Natural Selection, 1930), natural selection requires Mendelian inheritance to be effective. Genetic evolution works upon such a substrate and can therefore be Darwinian. Cultural (or memetic) change manifestly operates on the radically different substrate or Lamarckian inheritance, or the passage of acquired characters to subsequent generations. Whatever we invent in our lifetimes, we can pass on to our children by our writing and teaching. Evolutionists have long understood that Darwinism cannot operate effectively in systems of Lamarckian inheritance – for Lamarckian change has such a clear direction and permits evolution to proceed so rapidly that the much slower process of natural selection shrinks to insignificance before the Lamarckian juggernaut.

This crucial difference between biological and cultural evolution also undermines the self-proclaimed revolutionary pretensions of a much-publicised doctrine – 'evolutionary psychology' – that could be quite useful if proponents would change their propensity for cultism and ultra-Darwinian fealty for a healthy dose of modesty.[4] Humans are animals and the mind evolved; therefore, all curious people must support the quest for an evolutionary psychology. But the movement that has commandeered this name adopts a fatally restrictive view of the meaning and range of evolutionary explanation. 'Evolutionary psychology' has, in short, fallen into the same ultra-Darwinian trap that ensnared Daniel Dennett and his *confrères* – for disciples of this new art confine evolutionary accounts to the workings of natural selection and consequent adaptation for personal reproductive success.

Evolutionary psychology, as a putative science of human behaviour, itself evolved by 'descent with modification' from 1970s-style sociobiology. But the new species, like many children striving for independence, shuns its actual ancestry by taking a new name and exaggerating some genuine differences while ignoring the much larger amount of shared doctrine – all done, I assume, to avoid the odour of sociobiology's dubious political implications and speculative failures (amid some solid successes when based on interesting theory and firm data, mostly from nonhuman species).

Three major claims define the core commitments of evolutionary psychology; each embodies a considerable strength and a serious (in one case, fatal) weakness:

1 *Modularity*. Human behaviour and mental operations can be divided into a relatively discrete set of items, or mental organs. (In one prominent study, for example, authors designate a 'cheater detector' as a mental organ, since the ability to discern infidelity and other forms of prevarication can be so vital to Darwinian success – the adaptationist

rationale.) The argument for modularity flows, in part, from exciting work in neurobiology and cognitive science on localisation of function within the brain – as shown, for example, in the precise mapping, to different areas of the cerebral cortex, of mental operations formerly regarded as only arbitrarily divisible by social convention (production of vowels and consonants, for example, or the naming of animals and tools; see chapter by Annette Karmiloff-Smith).

Ironically, though, neurobiology and evolutionary psychology employ the concept of modularity for opposite theoretical purposes. Neurobiologists do so to stress the complexity of an integrated organ. Evolutionary psychology uses modularity to atomise behaviour into a priori, subjectively defined and poorly separated items (not known modules empirically demonstrated by neurological study), so that selective value and adaptive significance can be postulated for individual items as the ultra-Darwinian approach requires.

2 *Universality.* Evolutionary psychologists generally restrict their study to universal aspects of human behaviour and mentality, thereby explicitly avoiding the study of differences among individuals or groups. They argue that variations among individuals, and such groups as races and social classes, only reflect the influence of diverse environments upon a common biological heritage. In this sense, they argue, evolutionary psychology adopts a 'liberal' position in contrast with the conservative implications of most previous evolutionary arguments about behaviour, which viewed variation among individuals and groups as results of different, and largely unalterable, genetic constitutions.

I welcome much of this change, but, in one important respect, this new approach to universals and differences continues to follow the old strategy of finding an adaptationist narrative (often in the purely speculative or story-telling mode) to account for genetic differences built by natural selection. For the most publicised work in evolutionary psychology has centred on the universality in all human societies of a particular kind of difference: the putative evolutionary reasons for supposedly universal behavioural differences between males and females.

3 *Adaptation.* Evolutionary psychologists claim that they have reformed the old adaptationism of sociobiology into a new and exciting approach. They will no longer just assume, they now say, that all prominent and universal behaviours must, *ipso facto*, be adaptive to modern humans in boosting reproductive success. They recognise, instead, that many such behaviours may be tragically out of whack with the needs of modern life, and may even lead to our destruction – aggressivity in a nuclear age, for example.

Again, I applaud this development. If this principle were advanced in conjunction with the recognition that a putative evolutionary origin does not necessarily imply an adaptive value at all, then evolutionary psychology could make a substantial advance in applying Darwinian theory to human behaviour. But the advocates of evolutionary psychology proceed in the opposite direction by twisting the observation that the behaviour of modern humans may not necessarily have adaptive value into an even more dogmatic, and even less scientifically testable, panadaptionist claim. Evolutionary universals may not be adaptive now, they say, but such behaviours must have *arisen* as adaptations in the different ancestral environment of life as small bands of hunter-gatherers on the African savannahs – for evolutionary theory 'means' a search for adaptive origins.

The task of evolutionary psychology then turns into a speculative search for reasons why a behaviour that may harm us now must once have originated for adaptive purposes. To take an illustration proposed seriously by Robert Wright in *The Moral Animal*,[5] a sweet tooth leads to unhealthy obesity today but must have arisen as an adaptation. Wright therefore states, 'The classic example of an adaptation that has outlived its logic is the sweet tooth. Our fondness for sweetness was designed for an environment in which fruit existed but candy didn't.'

This statement ranks as pure guesswork in the cocktail party mode; Wright presents no neurological evidence of a brain module for sweetness and no palaeontological data about ancestral feeding. This 'just-so story' therefore cannot stand as a 'classic example of an adaptation' in any sense deserving the name of science.

Much of evolutionary psychology therefore devolves into a search for the so-called EEA, or 'environment of evolutionary adaptation', which allegedly prevailed in prehistoric times. Evolutionary psychologists have gained some sophistication in recognising that they need not postulate current utility to advance a Darwinian argument; but they have made their enterprise even less operational by placing their central postulate outside the primary definition of science – for claims about an EEA usually cannot be tested in principle but only subjected to speculation. At least an argument about modern utility can be tested by studying the current impact of a given feature upon reproductive success. Indeed, the disproof of many key sociobiological speculations about current utility pushed evolutionary psychology to the revised tactic of searching for an EEA instead.

But how can we possibly know in detail what small bands of hunter-gatherers did in Africa two million years ago? These ancestors left some tools and bones and palaeoanthroplogists can make some ingenious inferences from such evidence. But how can we possibly obtain the key information that would be required to show the validity of adaptive tales about an EEA: relations of kinship, social structures and sizes of groups, different activities of males and females, the roles of religion, symbolising, story-telling and a hundred other central aspects of human life that cannot be traced in fossils? We do not even know the original environment of our ancestors – did ancestral humans stay in one region or move about? How did environments vary through years and centuries?

In short, evolutionary psychology is as ultra-Darwinian as any previous behavioural theory in insisting upon adaptive reasons for origin as the key desideratum of the enterprise. But the chief strategy proposed by evolutionary psychologists for identifying adaptation is untestable and therefore unscientific. This central problem does not restrain leading disciples from indulging in reveries about the ubiquity of original adaptation as the source of revolutionary power for the putative new science. I detect not a shred of caution in this proclamation by Wright – embodying the three principal claims of evolutionary psychology as listed above:

> The thousands and thousands of genes that influence human behaviour – genes that build the brain and govern neurotransmitters and other hormones, thus defining our 'mental organs' [note the modularity claim] – are here for a reason. And the reason is that they goaded our ancestors into getting their genes into the next generation [the claim for adaptation in the EEA]. If the theory of natural selection is correct, then essentially everything about the human mind should be intelligible in these terms [the ultra-Darwinian faith in adaptationism]. The basic ways we feel about each other, the basic kinds of things we think about each other and say to each other [note the claim for universality], are with us today by virtue of their past contribution to genetic fitness.

Wright's closing sermon is more suitable to a Sunday pulpit than a work of science:

> The theory of natural selection is so elegant and powerful as to inspire a kind of faith in it – not *blind* faith, really . . . But faith nonetheless; there is a point after which one no longer entertains the possibility; of encountering some fact that would call the whole theory into question.
>
> I must admit to having reached this point. Natural selection has now been shown to plausibly account for so much about life in general and the

human mind in particular that I have little doubt that it can account for the rest.

This adaptationist premise is the fatal flaw of evolutionary psychology in its current form. The premise also seriously compromises – by turning a useful principle into a central dogma with asserted powers for nearly universal explanation – the most promising theory of evolutionary psychology: the recognition that differing Darwinian requirements for males and females imply distinct adaptive behaviours centred upon male advantage in spreading sperm as widely as possible (since a male need invest no energy in reproduction beyond a single ejaculation) and female strategies for extracting additional time and attention from males (in the form of parental care or supply of provisions, etc.). In most sexually reproducing species, males generate large numbers of 'cheap' sperm, while females make relatively few 'energetically expensive' eggs and then must invest much time and many resources in nurturing the next generation.

This principle of differential 'parental investment' makes Darwinian sense and probably does underlie some different, and broadly general, emotional propensities of human males and females. But contrary to claims in a recent deluge of magazine articles, parental investment will not explain the full panoply of supposed sexual differences so dear to pop psychology. For example, I do not believe that members of my gender are willing to rear babies only because clever females beguile us. A man may feel love for a baby because the infant looks so darling and dependent and because a father sees a bit of himself in his progeny. This feeling need not arise as a specifically selected Darwinian adaptation for my reproductive success, or as the result of a female ruse, culturally imposed. Direct adaptation represents only one mode of evolutionary origin. After all, I also have nipples not because I need them, but because women do and all humans share the same basic pathways of embryological development.

If evolutionary psychologists continue to push the theory of parental investment as a central dogma, they will eventually suffer the fate of the Freudians, who also had some good insights but failed spectacularly, and with serious harm imposed upon millions of people (women, for example, who were labelled as 'frigid' when they couldn't make an impossible physiological transition from clitoral to vaginal orgasm), because they elevated a limited guide into a rigid creed that became more of an untestable and unchangeable religion than a science.

Exclusive adaptationism suffers fatally from two broad classes of error, one external to Darwinian theory, the other internal. The external error arises from fundamental differences in principle and mechanism between, on the one hand, genetic Darwinian evolution and, on the other, human cultural change, which cannot be basically Darwinian at all. Since every participant in these debates, including Dennett and the evolutionary psychologists, agrees that much of human behaviour arises by culturally induced rather than genetically coded change, giving total authority to Darwinian explanation requires that culture also work in a Darwinian manner. (Dennett, as discussed earlier, makes such a claim for cultural change in arguing for the 'substrate neutrality' of natural selection.) But for two fundamental reasons (and a host of other factors), cultural change unfolds in virtual antithesis to Darwinian requirements.

First, topological: as the common metaphor proclaims, biological evolution builds a tree of life – a system based upon continuous diversification and separation. A lineage, after branching off from ancestors as a new species, attains an entirely independent evolutionary fate. Nature cannot make a new mammalian species by mixing 20 per cent dugong with 30 per cent rat and 50 per cent aardvark. But cultural change works largely by an opposite process of joining, or interconnection, of lineages.

Marco Polo visits China and returns with many of the customs and skills that later distinguish Italian culture. I speak English because my grandparents migrated to the United States. Moreover, this interdigitation implies that human cultural change needn't even follow genealogical lines – the most basic requirement of a Darwinian evolutionary process – for even the most distant cultural lineages can borrow from each other with ease. If we want a biological metaphor for cultural change, we should probably invoke infection rather than evolution.

Second, causal: as argued above, human cultural change operates fundamentally in the Lamarckian mode, while genetic evolution remains firmly Darwinian. Lamarckian processes are so labile, so directional and so rapid that they overwhelm Darwinian rates of change. Since Lamarckian and Darwinian systems work so differently, cultural change will receive only limited (and metaphorical) illumination from Darwinism.

The internal error of adaptationism arises from a failure to recognise that even the strictest operation of pure natural selection builds organisms full of non-adaptive parts and behaviours. Non-adaptations arise for many reasons in Darwinian systems, but consider only my favourite principle of 'spandrels'.

All organisms evolve as complex and interconnected wholes, not as loose alliances of separate parts, each independently optimised by natural selection. Any adaptive change must also generate, in addition, a set of spandrels or non-adaptive byproducts. These spandrels may later be 'co-opted' for a secondary use. But we would make an egregious logical error if we argued that these secondary uses explain the existence of a spandrel. I may realise some day that my favourite boomerang fits beautifully into the arched space of my living-room spandrel, but you would think me pretty silly if I argued that the spandrel exists to house the boomerang. Similarly, snails build their shells by winding a tube around an axis of coiling. This geometry of growth generates an empty cylindrical space, called an umbilicus, along the axis. A few species of snails use the umbilicus as a brooding chamber for storing eggs. But the umbilicus arose as a non-adaptive spandrel, not as an adaptation for reproduction. The overwhelming majority of snails do not use their umbilici for brooding, or for much of anything.

If any organ is, prima facie, replete with spandrels the human brain must be our finest candidate – thus making adaptationism a particularly dubious approach to human behaviour. I can adopt (indeed I do) the most conventional Darwinian argument for why the human brain evolved to a large size – and the non-adaptationist principle of spandrels may still dominate human nature. I am content to believe that the human brain became large by natural selection, and for adaptive reasons – that is, for some set of activities that our savannah ancestors could only perform with bigger brains.

Does this argument imply that all genetically and biologically based attributes of our universal human nature must therefore be adaptations? Of course not. Many, if not most, universal behaviours are probably spandrels, often co-opted later in human history for important secondary functions. The human brain is the most complicated device for reasoning and calculating, and for expressing emotion, ever evolved on earth. Natural selection made the human brain big, but most of our mental properties and potentials may be spandrels – that is, nonadaptive side consequences of building a device with such structural complexity. If I put a small computer (no match for a brain) in my factory, my adaptive reasons for so doing (to keep accounts and issue paycheques) represent a tiny subset of what the computer, by virtue of inherent structure, can do (factor-analyse my data on land snails, beat or tie anyone perpetually in tick-tack-toe). In pure numbers, the spandrels overwhelm the adaptations.

The human brain must be bursting with spandrels that establish central components of what we call human nature but that arose as non-adaptations, and therefore fall outside the compass of evolutionary psychology or any other ultra-Darwinian theory. The brain did not enlarge by natural selection so that we would be able to read or write.

Even such an eminently functional and universal institution as religion arose largely as a spandrel if we accept Freud's old and sensible argument that humans devised religious beliefs largely to accommodate the most terrifying fact that our large brains forced us to acknowledge: the inevitability of personal mortality. We can scarcely argue that the brain got large so that we would know we must die!

In summary, Darwin cut to the heart of nature by insisting so forcefully that 'natural selection has been the main, but not the exclusive, means of modification' – and that hard-line adaptationism could only represent a simplistic caricature and distortion of his theory. We live in a world of enormous complexity in organic design and diversity – a world where some features of organisms evolved by an algorithmic form of natural selection, some by an equally algorithmic theory of unselected neutrality, some by the vagaries of history's contingency, and some as byproducts of other processes. Why should such a complex and various world yield to one narrowly construed cause? Let us have a cast of cranes, some more important and general, others for particular things – but all subject to scientific understanding, and all working together in a comprehensible way. Why not admit for theory the same delight that Robert Louis Stevenson expressed for objects in his 'Happy Thought':

> The world is so full of a number of things,
> I'm sure we should all be as happy as kings.

Notes and References

1 Daniel Dennett, *Darwin's Dangerous Idea: Evolution and the Meanings of Life* (New York, Simon & Schuster, 1995).
2 Niles Eldredge, *Reinventing Darwin: The Great Debate at the High Table of Evolutionary Theory* (New York, John Wiley, 1995).
3 See my recent book *Life's Grandeur* (London, Cape, 1996, published in the USA as *Full House*, New York, Crown, 1996) for an account of this.
4 See such technical works as J. Barkow, L. Cosmides and J. Tooby (ed.), *The Adapted Mind: Evolutionary Psychology and the Generation of Culture* (Oxford, Oxford University Press, 1992); D. M. Buss, *The Evolution of Desire* (New York, Basic Books, 1994); and, especially, for its impact by good writing and egregiously simplistic argument, the popular book of Robert Wright, *The Moral Animal: Why We Are the Way We Are: The New Science of Evolutionary Psychology* (New York, Random House, 1994).
5 Ibid., *The Moral Animal*.

7

Colonising the Social Sciences?

Hilary Rose

At the turn of the century Darwin is everywhere. Marx has been pushed off his pedestal, though some hopefuls are convinced of his imminent reappearance. Freud is in trouble too; almost evicted from psychiatry, he clings on and arguably is even thriving in cultural studies and literary theory. The *New Yorker* sums it up neatly: biobabble, it asserts, is driving out psychobabble. As an indicator of this cultural shift it compares the past electioneering advice offered by pop psychologist Naomi Wolf to President Clinton, that he needed to project a more fatherly image, to her current advice to Al Gore to become more like an alpha male. But this either/or explanatory turn is part and parcel of our time; intellectual fashions shift in a constant kaleidoscope as competing theories, disciplines and gurus capture centre stage.

Today's kaleidoscope shakes down into three broad patterns and their protagonists. The first recognises and sees the case for the diversity of the disciplines; the second continues to insist on the hegemony of the natural sciences over all other form of knowledge; while the third shows an almost nostalgic longing to build the unity of the sciences. As part of this third strand we read John Tooby and Leda Cosmides as influential theorists of evolutionary psychology (EP), claiming that they want to establish this unity, not in the old and negative sense of reductive appeals to collapse all sciences to a lower, more basic level (typically physics), but in a more modest, contemporary sense in that scientific disciplines should not contradict each other.[1] In kindred spirit the founder of sociobiology and naturalist E. O. Wilson calls his latest book *Consilience*.[2] In this chapter I argue that although this respectful approach to other disciplines, both as agenda and title, is to be welcomed, the problem is that EP, like sociobiology before it, singularly fails to deliver. Beneath the new rhetoric lies the old project of colonising the social science under the banner of biology. Where its immediate progenitor sociobiology turned to animal behaviour, the new evolutionary psychology turns to a universal human nature as it evolved in the Pleistocene period.

The Origins of The Origin: *Darwin and Malthus*

Given that Darwin and evolutionary theory form the focal point of these recharged debates about the capacity of biology to singlehandedly explain human nature, it is fascinating to look back to the early nineteenth century, for this period, and above all the 1830s, was the crucible in which Darwin's theory of the origin of species was forged.[3] There has been no period since then until the present day in which such unconstrained hard Anglo-American capitalism has prevailed. The political economy of Reagan and Thatcher and their heirs has presided over both the destruction of the Victorian ideal of public service and the twentieth-century collectivist achievement of the welfare state. A restored and re-energised neoliberalism has provided the perfect ecological niche for a new wave of biology-as-destiny.

But let me go back to early nineteenth-century Britain. Then the intelligentsia, by contrast with today, were relatively few in number and in consequence, despite their personal commitment to, say, political economy, the novel, geology or the study of living nature, their reading practices ranged across all fields. What in the mid-twentieth century C. P. Snow was to call the two cultures were yet to appear; this earlier cultural elite pretty much shared a common culture. Charles Lyell's *Principles of Geology*, published in three volumes between 1830 and 1833, was of course carefully packed for the *Beagle* voyage. But the account of Darwin's leaving an important scientific text behind (no less than Humboldt), because he could not bear to set off without the latest Jane Austen, is both touching and revealing. The exchange worked both ways, for Charlotte Brontë's *Jane Eyre* is richly informed by phrenology, then the hottest materialist account of brain and mind. Darwin's diaries and notebooks show him keeping up with current intellectual concerns, drawing huge stimulus from his encounters with the theories and evidence from a diversity of fields and endlessly worrying about his always imminent loss of faith.

Thus we learn of Darwin in the Athenaeum in September 1838, reading a rapturous review of Auguste Comte's *Positive Philosophy*, feeling as one with the French mathematician in his determination to establish laws. For both Comte and Darwin law was the hallmark of a mature science. That Comte also coined the word sociology and was to become the father of positivistic sociology was of less moment; Darwin took what he wanted and left the rest – but not without distress. The Frenchman's atheist materialism intensifed Darwin's own rejection of free will, bringing his faith under even greater pressure. Reconciling faith and materialism was for Darwin not just an intellectual but an acute personal problem, for his beloved cousin Emma Wedgwood came from a deeply religious family. His marriage to her, and the constraints of their class, meant that unlike his socially less advantaged colleague Thomas Huxley he could never admit an explicit atheism.

However, it was Darwin's encounter later in the same month with Thomas Malthus's *Essay on the Principle of Population*, originally published in 1798, which was to prove decisive for *The Origin of Species*. Malthus's bleak message that the growth of human populations inexorably outstripped the available food supply was immensely widely read. Its thesis of the iron necessity of laissez-faire capitalism spoke directly both to the troubles of the times and to Darwin of a solution to his theoretical troubles.

The country was in deep economic recession. Terrible social suffering was accompanied by widespread riots by the urban poor and unemployed, particularly against the cruelties of the 1832 New Poor Law. In the great cities such as London and Liverpool the Poor Law Commissioners were unable to bring in the new laws for fear of producing even greater and potentially unmanageable social unrest. This protest was given political direction by the rise of the Chartist movement. Inspired by Christian socialism, the demands of the Chartists for social justice and universal suffrage mobilised hundreds and thousand of impoverished working-class men and women. With the Chartists at the comfortable bourgeois gates, Malthus's thesis that any help given to the poor did not just postpone but made the final reckoning even more terrible was enthusiastically received by those trying to contain or, even better, cut the Poor Law bills. Because Malthus's work was part of the new scientific political economy, it claimed law-like status for its propositions. This gave Malthus even greater legitimacy both with the reform-minded bourgeois Whigs[4] and the troubled biologist. The softening of the sixth edition, which Darwin read, hugely increased its acceptability, as in this version Malthus offers emigration of the surplus population as a kinder way out. The successful bourgeoisie is no longer compelled to witness the unsuccessful surplus population dying before their eyes; instead the poor, 400,000 a year, were loaded on to ships and exported to the colonies.

Central to understanding how this shared culture worked is the way in which Darwin uses the Malthusian theory of competition within human populations over scarce resources, and then more slowly explores and extends its utility in a biological context. Thus while Malthus divides his population between the innately Deserving Rich and the equately innately Undeserving Lower and Middling Classes, Darwin's populations are racialised. Travelling on the *Beagle* he had seen the destruction of the indigenous peoples by the settlers, those boatloads of surplus population endorsed by Malthusian law. He had witnessed and been revolted by General Rosa's genocide of the South American Indians and been torn by his feelings that the Tierra del Fuegans were both part of the human species and utterly disgusting. A racialised version of Malthus offered him the solution; the 'races' are in conflict for scarce resources and the best adapted will survive. 'When two races of men meet they act precisely like two species of animals – they fight, eat each other, bring diseases to each other &c, but then comes the more deadly struggle, namely which have the best fitted organisation, or instincts (ie intellect in man) to gain the day.'[5] For Darwin this best fitted 'race' was naturally the British.

Nonetheless Darwin's extreme sensitivity to the materialist implications of his theory of natural selection led him to postpone publication for almost twenty years. As Ted Benton reminds us, it was only the receipt of Alfred Russel Wallace's letter that prompted Darwin to publish *The Origin of Species* in 1859. By then Malthus was no longer in intellectual fashion, instead Victorian England had begun to set about improving the condition of her people; thus the context of publication was markedly different from the context of theorising. While energetic ideologues of laissez-faire were conspicuously present throughout the century, they no longer dominated; now the social interventionists were equally robustly present. These very different contexts have facilitated two very different readings of Darwin. Broadly speaking, the biologists' reading more or less erases Malthus and instead stresses Darwin's brilliant induction from meticulous observation and careful experiment. Equally broadly speaking, the social historians of science insist on the importance of social context. Many, as I have done here, insist on the relevance of Malthus and 1830s' laissez-faire for an adequate understanding of Darwinian theory. Much to the irritation of the biologists' reading of *The Origin* as a hugely innovative and purely scientific text, the social historians see Darwin's theorising as part and parcel of his times – the innovation lies in transferring a social theory into biological discourse.

Thus Anne Fausto-Sterling's chapter reminds us that Darwin's assumption of the automatic intellectual superiority of men was of a piece with the conventional values of men of his position and epoch. In

The Origin he begins the chapter on sexual selection with 'Man is more courageous, pugnacious and energetic than woman and has more inventive genius.' Such claims were not without criticism from contemporary feminists. But it is Antoinette Brown Blackwell's 1875 feminist rejoinder to Darwin's later book, *The Descent of Man* (1871), which creates a contemporary echo. She welcomes evolutionary theory but sees Darwin's rendering of it as androcentric. She argues (and it is a point echoed by feminist biologists, notably Ruth Hubbard[6] and Sarah Hrdy,[7] a century later) that Darwinian theory treats the evolution of human nature as if only men had evolved.

Fitness, Marx and Spencer

It is his contradictory use of the term 'fitness' which points to the problem. Typically when Darwin is discussing flora and fauna, fitness means reproductive success; however, when he is discussing human populations fitness suddenly no longer means reproductive success, as that would entail recognising the poor with their large families as the fittest. Suddenly 'fitness' becomes suffused with the dominant social values of his time, filled with the ideas of social progress and superiority that elsewhere are given no tolerance in Darwinian theory. This ambiguity around 'fitness' – not to say downright slipperiness – in the original Darwin texts means that this central concept sits there almost asking to be recruited around any political project, typically by the social conservatives but also by social revolutionaries.[8]

Thus the leading social theorists of the time, Herbert Spencer and Karl Marx, were both, though for very different reasons, immensely attracted by evolutionary theory. What particularly drew Spencer was the importance Darwin gave to competition as the mechanism of natural selection. It was not Darwin but Spencer, followed by the poet Tennyson, who put into cultural circulation the savage metaphor of 'Nature red in tooth and claw'. Spencer (like today's EP theorists) was primarily interested in the mechanism of competition and (again like the EP theorists) relatively uninterested in Darwin's grand project of providing an account of transmutation over time. Competition and natural selection served Spencer well enough in what was a fundamentally political project to explain why existing social hierarchies were natural and hence immutable. Despite Spencer's work being acclaimed as Social Darwinism, Darwin firmly dissociated himself from the theory; he regarded Spencerism as grandiosely abstract and empty of factual detail.

However, it was precisely the theory of change over time through conflict which attracted Marx and Engels. Preoccupied with explaining historical change, that is, the transmutation of societies (for example, from feudalism to capitalism), Marx and Engels had settled on class conflict as the key mechanism. That the great biologist also selected conflict as his key mechanism of change was seen as welcome reinforcement. In the Darwin family home, Down House in Kent, there is a copy of *Das Kapital* sent by Marx to the author of *The Origin*. Although it is difficult to imagine what the establishment Darwin's feelings were on receiving this tribute from the revolutionary and social pariah, it is clear from the mostly uncut pages that Darwin found Marx's text insufficiently compelling to struggle with his German.

Nonetheless, while Marx and Engels welcomed Darwinian theory, they simultaneously recognised this movement between social theory and biological theory. In a letter to Engels in 1862, Marx writes,

> It is remarkable how Darwin recognises among beasts and plants his English society with its division of labour, competition, opening up of new markets, 'inventions', and the Malthusian 'struggle for existence'. It is Hobbes's 'bellum omnium contra omnes' [the war of all against all] and one is reminded of Hegel's *Phenomenology* where civil society is described as a spiritual animal kingdom, while in Darwin the animal kingdom figures as civil society.[9]

These moves by social theorists of left and right to argue that their historical explanations were congruent with Darwinian theory could be seen as a chance for the unity of the sciences – which Tooby and Cosmides have advocated rather more recently. That Spencer and Marx saw themselves, regardless of how others saw them, as unquestionably practising science, facilitated that Victorian conviction that the sciences were coming of age. They were no longer contradicting one another but were achieving congruence across disciplines. What is fascinating is that Darwin will have none of it; he is much more concerned to claim the autonomy of biology. For although Darwin himself remained the great amateur, he is also part of the move – most visible in the third quarter of the nineteenth century – to professionalise biology in the universities.

The successful entry (and not so incidentally masculinisation) of biology as a new discipline within the universities alongside the old necessitated a defence of its autonomy and methods. Crucially biology could not be reduced either to chemistry or to physics. But the story of proliferating disciplines each claiming autonomy and distinctive methods has run ever since and shows no signs of diminishing. It would,

111

however, be a mistake to see the growth of new disciplines as one of simple linear growth. War, state policy and economic crises have played their part, as research always requires resources, but the twentieth century also saw the development of the research foundations which have at times played a highly interventionist role in shaping the direction of the disciplines.

Rockefeller Constructs Social Biology

One conspicuous shaper in the 1930s was the wealthy Rockefeller Foundation, which had enhancement of the rational control of human behaviour as its overall objective and saw the expansion of genetics as a key mechanism. Within this project the Foundation was concerned that the social sciences were becoming too autonomous and were no longer paying adequate attention to biology. The still new institution of the London School of Economics and Political Science (LSE) was seen as a key target and the US-based foundation offered the director, William Beveridge, a substantial grant to advance this scientific direction. Interwar universities were cash strapped and directors not in the business of looking gift horses in the mouth. Beveridge accepted the grant and in due course Lancelot Hogben, mathematician, biologist and Marxist, was appointed to the Chair in Social Biology. Even thirty years on at the LSE, the Hogben experience was described to novitiates joining the staff as one of near academic disaster for the UK's major institution dedicated to the social sciences. Whichever senior academic told the story the punchline was constant, 'There were even toads hopping out of his room into the corridor.' That Hogben was one of the first on the British left to spot the inexorable and fascist direction of eugenics and became one of the most powerful voices against the new trend in the 1930s is a happy irony.

Unsurprisingly, despite the biologistic foundations of much early social theory, social scientists wished to preserve their newly established autonomy, although they were entirely happy to contribute to discussions of evolutionary theory. Indeed, as the LSE sociologist and Darwin scholar Donald Macrae observed, social and biological theorists contributed almost equally in 1909 at the fifty-year celebration of *The Origin*, but by the centenary celebration Macrae was the only social theorist present.[10] Despite this, when the suggestion was made that the study of the social life of red deer should be added to the sociological curriculum the Darwin scholar was outraged – a scene still vivid in my memory from being a junior academic at the LSE in the 1960s. While sociology undergraduates were taught something of ethology as an adjacent field, this did not mean that social sciences were about to subordinate their explanations to those of the life sciences.

The hostility was only modified by the appointment of John Ashworth, biologist and former Chief Scientific Adviser to the government, to the directorship of the LSE in 1990. Ashworth clearly wanted to bring about a unity of the sciences – at least between the social sciences and biology. To this end he established the Centre for the Philosophy of the Social and Natural Sciences and began bringing people in to develop the new field of evolutionary psychology. One of the appointments was that of philosopher turned sociobiologist Helena Cronin, author of *The Ant and the Peacock*. Cronin set up the Darwin@LSE address, ran the highly media-visible Darwin seminar programme and edits the series of booklets *Darwinism Today*. Ashworth's hopes for this new venture were evident:

> More recent attempts by E. O. Wilson in the 70s, under the banner of sociobiology, were also not encouraging despite there being theories of genetics and the mechanism whereby seeming altruistic behaviour might have been selected by evolutionary process. Wilson's assertive tone was too reminiscent of Spencer's and many reacted, not to the new insights that Wilson was popularizing but to the previous attempts of Spencer and the social biologists of the thirties . . . Most biologists accepting the charges of insensitivity and tactlessness kept their heads down and few will now wish to be called sociobiologists. But in such guises as evolutionary psychology . . . accessible and sensitive accounts of the results have been published in a form that might now lead to something other than a dialogue of the deaf.[11]

But what was strange for an ostensibly academic and scientific innovation was that Ashworth's text formed the introduction to a special issue of the centre-left magazine, *Demos*, highly influential in New Labour circles. If EP was indeed an academic subject, even a science, why choose a political magazine rather than an academic journal as a launch pad? Although Ashworth wanted to establish what he saw as the new approach of EP, he failed to recognise that insofar as this was new, the newness was in the main a US phenomenon. Apart from a very small handful of UK psychologists committed to neo-Darwinism, the new LSE centre had largely to make do with sociobiologists, evolutionary psychologists' close kin. In contrast with Ashworth's high hopes for EP, Richard Dawkins, as the single most elegant and influential exponent of sociobiology in the UK, is refreshingly blunt. Setting to one side the contribution of the psychologists, he describes EP as 'rebranded sociobiology'.[12]

Darwin@ LSE attempted to marry sociobiology and evolutionary psychology into one indissoluble unity which took evolutionary theory, above all adaptationism, as a universal and irrefutable explanation. Unequivocal sociobiologists from Richard Dawkins and Robert Trivers to Matt Ridley and Kingsley Browne dominated the seminar series, and were joined by the new voices from evolutionary psychology, most visibly the high-profile MIT cognitive psychologist Steven Pinker, but also the Canadian-based psychologists Martin Daly and Margo Wilson.

The Rhetoric of Arrival

In 1996 the LSE appointed the distinguished sociologist Tony Giddens to the directorship. The environment has inevitably changed – not least for the Centre. Today Darwin@ LSE continues as an email address, but new staff with a background in psychology have been brought in and the highly visible seminars have been discontinued. The relevance of evolutionary theory to the social and natural sciences is still to be explored within the LSE.

Ashworth's optimistic claim that evolutionary psychology has arrived as a science, leaving aside the unfortunate Spencerian overtones of sociobiology, is interesting as a rhetorical move. Such distancing of a claimed new science from its immediate ancestor is integral to the peculiar history of the story of 'biology-as-destiny' throughout the twentieth century. Each new wave of biology-as-destiny distances itself from its precursor, which is portrayed as too populist, or, worse still, must now be dismissed as having admitted an unequivocally political agenda into its discourse. By contrast, whatever the present wave, it has now achieved full scientific status. This rhetoric of arrival, which is built into the twists and turns of biology-as-destiny, is radically different from that of normal science. Take the conception of the Human Genome Project, that moment in the mid-1980s when biology entered big and expensive science. There were lots of big promises and some sharp arguments among the biological elite about the biomedical and cultural implications in the leading US journal *Science*, but at no time was it thought necessary to say that this, unlike its predecessors, is now science. Its status was self-evident to friend and foe alike.

Nowhere is this peculiar distancing better shown than in the narrative sociobiology tells about its own prehistory. Precursor texts such as those of Robert Ardrey, *The Territorial Imperative* (1966), Konrad Lorenz, *On*

Aggression (1966) Desmond Morris, *The Naked Ape* (1967), Robin Fox and Lionel Tiger, *The Imperial Animal* (1970), are essentially Hobbesian in their view of human nature as inexorably nasty, aggressive and territorial, and thoroughly Spencerian in their 'Nature red in tooth and claw'. (No space for Kropotkinist co-operation in this version of evolution).[13] The intellectual vulnerability of these texts to criticism from within and without biology, and perhaps most of all Stephen Jay Gould's lethal characterisation of the entire genre as 'pop ethology', goes a good way to explain why E. O. Wilson distanced himself from these precursors to speak contemptuously of them as 'mere advocates'. Now, he argued, with the publication in the 1970s of *On Human Nature*[14] and *Sociobiology: The New Synthesis*,[15] sociobiology has arrived as a science.

Science as the View from Nowhere?

Wilson's texts (setting aside the sharp political and scientific controversy to which they gave rise) constantly insist on their scientificity. His account of human nature, he claims, exactly coincides with that classical objectivist stance of science as the 'view from nowhere'. This gives rise to two sets of difficulties. First, Wilson cannot even persuade biologists such as Ashworth, who are friendly to sociobiology's core project and hope for the unity of the life and social sciences, that what he does is quite science. Second, this view of science as privileged knowledge, outside cultural, economic and social influence, has come under intense interrogation over the last three decades from the philosophy, history and sociology of the sciences. This interrogation has not been received without fierce resistance from sections of the natural science community. The high tide of this highly charged debate concerning the cultural status of both science and its critics was unquestionably the publication of *The Higher Superstition*,[16] attacking the new social studies of science, followed shortly by the counter-publication, *Science Wars*.[17] Academically weak as *The Higher Superstition* was,[18] it touched a chord in the natural science community, many of whom were feeling that non-scientists no longer deferred to the authority of science and scientists in the way they had in the past.

In this situation the old appeal to scientificity by those proclaiming versions of biology-as-destiny no longer carries such cultural weight. They may make the historic cultural assertion that as scientists they are true Baconians,[19] merely holding up a mirror to nature, but increasingly a good part of their audience suspects that they begin by making that selfsame mirror in the image of existing social relations. The sociobiologist David Barash's rhetorical appeal in defence of his mysogenist claims that men are naturally predisposed to rape, 'If Nature is sexist don't blame her sons,'[20] can no longer plug into the old deference to science as the view from nowhere.

Without necessarily buying into a full postmodernist agenda and declaring the death of truth, the public understanding of science at the turn of the century is more sceptical, more questioning, than ever before. Such scepticism has been fostered by the hostility of two powerful social movements deeply concerned with how science represents and acts upon both human and green nature. Feminism has challenged biological claims that women's subordinate place can be explained in terms of their hormones, brain size, reproductive capacity, and so on, and has fought for more reliable accounts of women's biology. Environmentalism has seen the growing threat to the environment and challenged science's reductionism and exploitative stance to nature. At the same time, historians, philosophers and sociologists often influenced by these two movements have developed both micro and macro accounts of scientific knowledge, showing science as socially constructed. These critiques have been widely taken up in media accounts of science. Science writers may stay faithful and defer to science but the cultural commentators and the news reporters have long given up on what they increasingly see as false deference. After Chernobyl, BSE and GM food the public is no longer content to accept the reassurance of industrial and governmental scientists. Their view is seen as partial in two senses. First, because their reductionist training hinders them from seeing and understanding the whole human ecosystem, although it is within this that risk takes place; and, second, because neoliberal governments and industry are more interested in wealth creation than in quality of life.

Today's public discussion of science is not only attentive to who pays for research and its commercial or governmental links. In a more subtle way the social, economic and cultural interests that come into the research as part of the researcher's taken-for-granted values come under question. Are these interests going to be problematised and dealt with as part of the research, or are they to be left unaddressed, with the scientists simply huffing about their objectivity and integrity? Unsurprisingly huffing as an alternative to dealing transparently with the question of interest is increasingly unconvincing either to fellow scientists or to the general public. Everywhere in cultural and social life democratic societies increasingly assume that our educational, cultural, legal and political institutions work better if they reflect the diversity of the population they serve. When science is included amongst these institutions, as it increasingly is, the development carries some interesting epistemological implications.[21]

EP on Different Sides of the Pond

In the UK evolutionary psychology has needed the alliance with sociobiology, an arrangement with which sociobiology is entirely happy. However, in the United States there are enough psychologists involved in the construction of evolutionary psychology for it to have become widely accepted as a new academic discipline. Further, with some conspicuous exceptions (not least Gould), in its country of origin EP has not received the same intensity of intellectual criticism as in the UK. Thus Pinker's book *How the Mind Works* was hostilely reviewed in the UK in the two leading literary weeklies, but was, by and large, well received in the USA. While a handful of EP theorists have highly visible media profiles, I want to focus on the rather less publicly visible John Tooby and Leda Cosmides (T&C) as the key theorists who have led this innovation within psychology. Pinker, for example, says, 'I have many intellectual debts . . . but most of all to John Tooby and Leda Cosmides.'[22]

Their much-cited chapter 'The Psychological Foundations of Culture'[23] claims that evolutionary theory explains, through individual psychology, nothing less than universal human nature as laid down in the Pleistocene. It must be said that evolutionary psychology represents an epistemological break with the biology-as-destiny narrative of psychometrics, which claims that every feature of human psychology can be quantified. Where in the case of race and IQ theory, from, say, Hans Eysenck and Arthur Jensen in the 1970s to the most recent restatement in *The Bell Curve*, the category of race is fixed in nature, Tooby and Cosmides reject the concept entirely. Like mainstream biology and, above all, population genetics, they insist on the unity of the human species. Indeed they go further as they eschew not only race and IQ theory but the whole of the genetics of behaviour. Attention to difference, whether by behaviour geneticists or social scientists, spoils their assumption of a universal, gendered human nature fixed in the Pleistocene. Hence they concentrate on what they claim are universal cultural practices, such as the age difference between men and women in marriage, child abuse by stepfathers, universal standards of female beauty, and so forth, arguing that these are evolutionary adaptations. The social sciences, they insist, must conform to Pleistocene psychology. 'The design of the human psychological architecture structures the nature of social interactions humans can enter into, as well as the selectively contagious transmission of representations between individuals.'[24] As another EP theorist, Laura Betzig, puts it, 'From pregnancy complications, to the stress response, to the beauty in symmetry, to the attraction of money, to the historical tendency of the rich to favor firstborn sons, everything we think, feel and do might be better understood as a means to the spread of our own – or of our ancestors' – genes.'[25]

For those conscious that scholars of prehistory work with highly fragmentary evidence, from shards of bones, fossils and very occasionally entire bodies preserved by ice or some geological quirk, the belief that late twentieth-century people can know the human psychological architecture of our early ancestors with any degree of certainty and accuracy is difficult to take seriously. There is an ongoing technical debate as to whether even the sex of that early hominid, named 'Lucy' within hours of discovery, was quite so clear cut as the discoverers claimed and the media amplified.[26] Was Lucy really a she, or a small Dr Pepper he? It is evident that viewing prehistoric people through the sexually dimorphic prism of late twentieth-century embodiment has inbuilt dangers. The rhetoric of 'reverse engineering' which EP deploys to explain its search for the present in the human psychology of prehistory is an inadequate cover-up for a fundamental problem of evidence and its interpretation. If Lucy's sex is a matter for technical debate, claims of certainty about prehistoric innate psyches look extraordinarily thin. No wonder many social scientists looking at the construction of the family produced by EP see them as embarrassingly like the Flintstones. The trouble is that we are meant to giggle at this Stone Age family with the gender mores of 1950s America but EP asks us to take its account of gender relations seriously.

Setting Up a Straw Enemy – the SSSM

It is important to understand how T&C propose to achieve their goal of establishing the unity of the sciences. They first set themselves the task of destroying the claims of the social sciences to be able to explain distinctively social phenomena. To this end they construct something they call the SSSM – the standard social science model – which, they claim, dominates the social sciences and prevents them from taking their place within the unity of the sciences. They see the SSSM as setting aside any biologically founded theory of human nature. Arbitrarily T&C exclude both economics and political science from their SSSM model, which is rather like excluding physiology and biochemistry from an account of the life sciences. Indeed, the exclusion is rather more serious, as economics has long had a model of economic man (*sic*), and political science, above all as political theory, has been one of the most fertile sources of theories of human nature. Hobbes, Rousseau, Locke and Marx, to name a few, have been neither silent nor without influence. Indeed, given that a theory of human nature has historically been seen as the starting point of political theory, it would seem the discipline to focus on, not to exclude in this arbitrary way.

T&C indict the social sciences, 'particularly sociology and anthropology', as deliberately cutting themselves off from the natural sciences. Given the scale and intensity of the current academic effort by social scientists and humanities scholars, including both anthropology and sociology, to study the natural sciences and technology, this seems a strange claim. At the four-yearly joint meetings of the 4S (Social Study of Science in Society) and EASST (European Association for the Study of Science and Technology) there are several thousand participants drawn from all over the world. For that matter, leading social theorists, such as the LSE-based Giddens along with Ulrich Beck and Scott Lash, have developed the concept of 'the risk society' in order to grapple with the contemporary phenomenon of scientific and technologically produced risk. Feminist work on reproduction and the new technologies of assisted conception has long ago overcome Descartes' dualistic error. Thus, as anthropologist Marilyn Strathern put it so memorably, a baby is simultaneously biological and social, it is not either/or. Indeed, as Tom Shakespeare and Mark Erickson's chapter makes clear, so is disability, and Tim Ingold's chapter, so is walking.

It is evident that sections of the natural science community find themselves uncomfortable and, in the case of some (not least Gross and Levitt), completely enraged by this academic claim from outside natural science to have something to say about science's construction of knowledge. Nonetheless the existence of this academic development cannot simply be denied. Yet it seems that T&C's highly scientistic version of psychology has prevented them from being able to see this development. In consequence the EP theorists sound dangerously close to the classic British newpaper headline: 'Fog in Channel! Continent Cut Off!'

Suicide and Love

To argue their case, Tooby and Cosmides focus on two figures, Emile Durkheim for sociology and Clifford Geertz for anthropology, somewhat arbitrarily selected from an entire century of the two disciplines. The former was one of the founding figures of sociology in the late nineteenth century, the latter is a current, highly influential US cultural anthropologist. Here T&C's straw model of the SSSM demonstrates its flexibility. Where in the case of Durkeimian sociology the model is flawed because it postulates the social as not reducible to another level, Geertzian anthropology is flawed because it argues that the cultural is not reducible. Thus the central concept of the SSSM slips opportunistically between the 'social' and the 'cultural' according to the discipline under criticism.

Culture, T&C say, is 'the protean agent that causes everything that needs explaining in the social sciences', then, recognising that this might be a claim too far, they retreat from this sweeping statement with 'apart from those few things that can be explained by content-general psychological laws, a few drives and whatever superorganic processes (e.g. history, social conflict, economics) that are used to explain the specificities of a particular culture.'[27] Their retreat gives the weakness of their denunciation of culture away, for what is excluded by their biologistic language of 'superorganic processes' is, despite the brackets, the very stuff from which social theory is built.

Tooby and Cosmides discuss just one book in Durkheim's entire corpus, *The Rules of Sociological Method* (1895), but they do not indicate the curious circumstances surrounding its publication, for Durkheim wrote the book as a tactic to secure a chair for sociology at the Sorbonne against the opposition of the psychologists. (*Plus ça change?*) Of a piece with the new experimental biology, defending its autonomy and irreducibility to chemistry and physics, Durkheim defended the autonomy of the social as a legitimate level of analysis irreducible to individual psychology. This argument is most elegantly shown in his classic study *Suicide*.[28] He takes this phenomenon, which seems above all to be about private agony and individual psychology, to demonstrate the autonomy of the social. Using pioneering statistical techniques (today, while admiring the innovation, we see them as slightly dodgy period pieces), he showed that personal agony and suicidal outcome were causally linked to religious values which bind or fail to bind the individual to the group and thence to life. Durkheim's theory thus explains both why Protestants kill themselves and why Catholics in the selfsame agony do not. But rather than engaging with the evidence for social explanations which the great French positivist marshals, T&C simply dismiss the possibility of the autonomy of the social in favour of reducing explanation to the level of individual psychology.

What is truly remarkable about Tooby and Cosmides is that they dismiss this classic and theoretically satisfying work but enthuse about the unsatisfying explanation of sexual abuse and violence by stepfathers proposed by psychologists and animal behaviourists Martin Daly and Margo Wilson.[29] Even though the evidence has been clear for a number of years that children in households where the mother is living with a man other than the biological father are at greater risk, explanations have been difficult to establish. But, as with Durkheim's explanation of suicide, an adequate theory must be able to explain both why a small minority kills and/or abuses and also why the majority do not. Here Daly and Wilson can only equivocate. First, working from a number of animal studies in which new male langur monkey and mouse partners kill the infants of earlier males, they proceed to argue that human parental love is genetic. (Incidentally they get no support from the primatologist Sarah Hrdy, whose work on langur monkey infanticide is key to their thesis. In the *New Scientist* she is recently quoted as saying, 'Human violence towards babies and infants may be tragic but it's nothing like what a langur male is doing.'[30]) Acknowledging that their genetic love thesis causes them some difficulty in explaining why the majority of stepfathers *do not* kill and abuse, and why a minority of genetic fathers *do*, Daly and Wilson then urge at various points in their argument, (i) that their results should be treated with caution, (ii) that they are not making an adaptationist explanation, but (iii) that theirs is a Darwinian explanation. These three are hard to reconcile.

All they can do is to appeal to the sort of free will commonly invoked by evolutionary psychologists, not least Pinker, either when the predicted adaptationist outcome does not happen or when they find the outcome ethically distasteful. Such a reliance on free will is of course significantly anti-Darwinian, for while Darwin has plenty of trouble with faith he is clear from very early on that free will has no part in his schema.

To make their universalistic case of the Cinderella effect[31] and their model of genetic parental love, Daly and Wilson marshal a large body of criminological data and draw on a number of human studies from a variety of contexts. Their analysis demonstrates two weaknesses: first, they statistically homogenise what are potentially quite different household structures (and assume that household equals family); and, second, they avoid powerful evidence that does not support their case. Thus they first conflate orphans like the original Cinderella with the contemporary child, who is more likely to have a living genetic father and a new live-in male partner of her mother – who may or may not be trying to become her stepfather. Indeed most UK press reports covering cases of violence and abuse analytically and linguistically distinguish between describing the male perpetrator as a stepfather or as the woman's lover. It is embarassing to have to praise the journalists' account of family and sexual relationships as more precise than that of the psychologists.

Even where a couple are attempting to build a second family, this is typically associated with psychological strain between the first family and the second, compounded by financial pressures. Rather obvious matters of context explain better why some men ill-treat their partners' children.

Because Daly and Wilson assume genetic parental love they have serious difficulties in explaining the lesser levels of abuse and violence in the case of adopted children. For where the parenting intentionality of both adults has been carefully assessed, this social selection protects children even better than shared genes. What price a Darwinian explanation for the greater love of adoptive parents? Indeed, if we chose to go down a policy road based on research aiming to reduce violence and abuse against children, the adoption evidence would surely suggest increasing social selection. Natural selection is self-evidently a non-explainer. As an interim step we would need to gather data on the parenting perfomance of parents who have received technological assistance with conception, for these too are assessed (though much less thoroughly) on their parenting potential. With all the evidence in, and assuming that those who have had to make that much more effort to conceive do perform better as parents, we would then be ready to assess prospective natural parents to ensure high-quality parenting. A truly energetic policy of social selection for parenthood could follow and violence and abuse diminish. As supporters of human rights with doubts about scientist eugenicist fantasies, we might well decide not to buy into this scenario, but in terms of the evidence social selection has a great deal more going for it than Daly and Wilson's rendition of natural selection.

But it is not necessary to criticise Daly and Wilson, for, as John Horgan[32] points out, Steven Pinker as their very public admirer has, with his own adaptationist theorising on infanticide, entirely shredded any credibility the Wilson–Daly thesis had. First, his admiration. When a number of fashionable intellectuals were asked in 1997 by the *New York Times Magazine* to name a book they had read twice, Pinker named Daly and Wilson's *Homicide*.[33] However, he subsequently wrote a piece for the *New York Times Magazine*[34] in which he discussed women killing their newborn babies and delivered himself of the view that such an act, where resources were minimal, could be an adaptationist response. He argued that the psychological module which normally produced protectiveness in mothers for their newborns might be switched off by the challenge of an impoverished environment. At this point evolutionary adaptationist reasoning becomes an absurd Catch-22 proposition: both killing and protecting are explained by evolutionary selection. Used like this selection explains everything and therefore nothing.

Furthermore, Daly and Wilson's thesis of parental genetic love has tremendous difficulties if they take on board two episodes of mass genetic parental cruelty. These throw into question their biologised explanation of protective love. The first lies in the recent past – the US history of slavery. The institutionalised rape of African women slaves by their white male owners and overseers was, as the historical record makes transparent, routinely associated with a brutal indifference to the fate of their genetic offspring. The second is still current. The high levels of foetal sexual selection, abortion and female infanticide in India because of the insistence of genetic fathers on male children is today, despite legislation, significantly skewing that country's population. China has similar troubles. If parental love is genetic and so compelling, how come racism and patriarchy can overcome it on such a huge scale?

Nature verses Nurture?

Perhaps the most perverse of Tooby and Cosmides's assertions is that sociological and anthropological adherents to the SSSM have been preoccupied by the nature/nurture debate. Now while it is the case that some of the earliest sociologists and anthropologists were, like Darwin and other men of their time, both ethnocentric and androcentric, sociology and anthropology have not been much interested in the heritability question over the last fifty years. Instead the nature/nurture debate has been the preoccupation of what we may call 'Anglo-Saxon' psychometrics. Ever since Galton, psychometricians have sought to measure what he so memorably called hereditary genius. But whilst behaviour geneticists have continued to pursue genius, now rebranded as 'g' or general intelligence, and are working to locate polymorphisms which correspond with these measures, sociologists in particular have taken the view that the notion of innate intelligence is unresearchable.

IQ scores are far too culturally specific to be used in the way the psychometrists propose.

Further, T&C see SSSM adherents in the most surprising places. For example, they see the evolutionary palaeontologist Gould, the population geneticist Richard Lewontin, the neurobiologist Steven Rose and the psychologist Leo Kamin as loyal to the SSSM, and claim that such loyalty assures the biologists that their discipline is 'intrinsically disconnected from human affairs'. This is a perverse reading of, say, *The Mismeasure of Man*[35] and *Not in Our Genes*.[36] In both books the authors are concerned to show how the preoccupations of the social context have shaped the biological accounts and frequently led to the production of poor and biased science. Lewontin, Rose and Kamin also offer an alternative theory to biological determinism more robust than the rather weak concept of interaction between nature and nurture, proposing a dialectical relationship between the organism and the environment. Picking up on the idea of the active organism, though deeming the Hegelian language of dialectic unnecessary, Karl Popper argued for what he called active Darwinism. But T&C have no space for an active evolutionary theory, they long for a world of unchanging categories. It is not by chance that the metaphors which occur again and again in their writing are those of fixity; they speak of 'architecture', 'mechanisms' and 'information system'; they have no space for language which emphasises process, like 'open systems', 'active Darwinism', 'fluid genomes' or even 'development'. But then EP, like Spencer's original Social Darwinism, does not share Darwin's preoccupation with transmutation over time.

Fundamentalist Diagnostics and Flexible Politics

The sharp dichotomy that evolutionary psychology makes between the social/cultural and the biological approach precludes the possibility that even now, in the context of the media-hyped science wars and the fragmented knowledges of the academy, multi-disciplinary conversations can and do take place. Those who engaged in the multi-disciplinary and synergistic seminar that produced this book are all too conscious of the many pleasures and occasional difficulties of such conversations. It is not as easy a task as it was for Darwin and his contemporaries, but it is still possible to demonstrate that we are able to listen to the diversity of field and approach without either rushing to colonise or to make grandiose universalisms, such as that unchanged and unchanging Pleistocene genes are us.

For social theorists worrying about problems of structure and embodied agency (the social theorist's contextualised version of free will), the cultural success of this latest version of biology-as-destiny simply adds to the analytic problems. What can only be characterised as the new fundamentalist Darwinism looks to claim its place among the other contemporary religious certainties of fundamentalist Islam, Christianity and Judaism, offering certainty in uncertain and troubled times. In today's pick-and-mix culture, evolutionary psychology proposes itself as yet another possible opiate for the people. Freud and Marx are dead – so long live Darwin.

Yet like the religious fundamentalists, the fundamentalist Darwinians who wish to colonise the social sciences have political as well as cultural objectives. Under the banner of an evolutionary account of human nature the new discipline offers to guide society. Some, like T&C, are cautious: 'it will be a long time before scientific knowledge of the aggregation of mechanisms that comprise the human psychological architecture is reliable enough and comprehensive enough to provide the basis for confident guidance in matters of social concern'.[37]

Other EP advocates are more ambitious and want to set EP's theory of human nature at the heart of their political programmes now. But what is fascinating is the contrast between the authoritarian certainty about Pleistocene psychology and the flexibility of the political projects. Thus animal rights philosopher Peter Singer wants to construct a *Darwinian Left*.[38] Right-wing libertarian Matt Ridley[39] sees Darwinian theory as pointing to the desirability of dismantling the welfare state, including the NHS, as these institutions inhibit virtue. Helena Cronin,[40] by contrast, sees EP as having important messages for feminists, as natural difference must be invoked to secure gender justice in the courts, while the cultural historian Francis Fukuyama in *The End of Social Order*[41] seems more concerned with the fate of men. Thus he sees male unemployment as disturbing the process of natural selection and therefore wants the state to intervene on behalf of Darwinian theory.

No wonder those who remain unconverted by the revealed truths of EP are made uneasy. And it is equally no wonder that some, exhausted by the difficulties of confronting our uncertain and sometimes ugly world, turn with an almost audible sigh of relief to such fundamentalist diagnostics and pick-your-own politics. Meanwhile attempting to explain genocidic conflict, globalisation, the ecological crisis, mass rape as a weapon of war, famine and disaster, new infectious diseases or the growing gap between rich and poor requires an array of analytic tools from many disciplines. Confronting such horrors and finding a political route beyond them requires both social courage and imagination. In this situation giving up responsibility for grappling with cultural and social complexity and embracing facile evolutionary universalisms is a moral and intellectual cop-out.

Notes and References

1 John Tooby and Leda Cosmides, 'The Psychological Foundations of Culture', in Jerome Barkow, Leda Cosmides and John Tooby (ed.), *The Adapted Mind: Evolutionary Psychology and the Generation of Culture* (New York, Oxford University Press, 1992).

2 E. O. Wilson, *Consilience: The Unity of Knowledge* (London, Little, Brown, 1998).

3 This section owes a debt to a number of Darwin biographies, particularly that of Adrian Desmond and James Moore, *Darwin* (Harmondsworth, Penguin, 1992), p. 267. R. M. Young's work on Darwin's metaphor has been influential in forming this view. 'Darwinism is Social', in D. Kohn (ed.), *The Darwinian Heritage* (Princeton, NJ, Princeton University Press, 1985).

4 The two dominating political parties then were the Whigs and the Tories. The former were the reformers, supporting free competition, religious toleration and a gradual extension of suffrage – the precursors of the Liberals. The latter were the party of conservatism, ranging from the landed gentry to the Church of England.

5 Darwin's 1839 notebooks, cited by Desmond and Moore, *Darwin*, p. 267.

6 Ruth Hubbard, 'Have Only Men Evolved?', in R. Hubbard, M. S. Hennifin and B. Fried, *Women Look at Biology Looking at Women* (Cambridge, MA, Schenckman, 1979).

7 Sarah Blaffer Hrdy, *The Woman That Never Evolved* (Cambridge, MA, Harvard, 1981).

8 Darwin has also been recruited around a peace biology. Paul Crook, *Darwinism, War and History* (Cambridge, Cambridge University Press, 1994).

9 Karl Marx and Friedrich Engels, *Selected Correspondence* 1846–1895 (London, Lawrence and Wishart, 1943), 9, pp. 125–6.

10 D. G. Macrae, 'Darwinism, the Social Sciences and Social Evolution', *Ideology and Society* (London, Heinemann, 1961).

11 John Ashworth, 'Introduction', *Demos*, special evolutionary psychology issue (October 1996), p. 3.

12 Interview with *The Evolutionist*, transcribed on the Evo-Psych website (November 1999).

13 Peter Kropotkin's classic anarchist text *Mutual Aid* (1898) saw co-operation as crucial in both human society and also among plants and animals. *Mutual Aid* is the forerunner of contemporary ideas of the human ecosystem.

14 E. O. Wilson, *On Human Nature* (Cambridge, MA, Harvard University Press, 1978).

15 E. O. Wilson, *Sociobiology: The New Synthesis* (Cambridge, MA, Harvard University Press, 1975).

16 Paul Gross and Norman Levitt, *Higher Superstition: The Academic Left and Its Quarrels with Science* (Baltimore, John Hopkins University Press, 1994).

17 Andrew Ross (ed.), *Science Wars* (Durham, Duke University Press, 1996).

18 The most arrogant and ill-infomed moment in the *Higher Superstition* is surely when Gross and Levitt propose that their colleagues who teach English and who evidently lack respect for science should be fired. Their certainty that as natural scientists they could then lash up an entirely satisfactory English course takes the biscuit!

19 Wilson chooses a quote from Bacon for the title page of *Consilience*.

20 David Barash, *The Whispering Within* (New York, Harper & Row, 1979), p. 55.

21 My personal conviction is that the numerical change in the gender and ethnic composition of the labour force in an age influenced by feminism and human rights will bring different concerns and values into science.

22 Steven Pinker, *How the Mind Works* (Harmondsworth, Allen Lane, 1998), p. x.

23 Tooby and Cosmides, in *The Adapted Mind*.

24 Ibid., p. 48.

25 Laura Betzig (ed.), 'Introduction: People are Animals', *Human Nature: A Critical Reader* (New York, Oxford University Press, 1997), p. 2.

26 Lori Hager, 'Sex and Gender in Paleoanthropology', in L. Hager (ed.), *Women in Human Evolution* (London and New York, Routledge, 1997).

27 Tooby and Cosmides, in *The Adapted Mind*, p. 48.

28 Emile Durkheim, *Suicide* (1897; London, Routledge & Kegan Paul, 1952).

29 Martin Daly and Margo Wilson, *Homicide* (New York, Aldine de Gruyter Hawthorne, 1988) and 'Evolutionary Social Psychology and Family Homicide', *Science* (28 October 1988), pp. 519–24.

30 Interview with Sarah Blaffer Hrdy, *New Scientist* (11 December 1999), p. 44. See also Sarah Blaffer Hrdy, *Mother Nature: Natural Selection and the Female of the Species* (London, Chatto & Windus, 1999).

31 Martin Daly and Margo Wilson, *The Truth about Cinderella: A Darwinian View of Parental Love* (London, Weidenfeld & Nicolson, 1998).

32 John Horgan, *The Undiscovered Mind: How the Brain Defies Explanation* (London, Weidenfeld & Nicolson, 1999), p. 184.

33 Steven Pinker, 'Takes Twice, Indeed Thrice', *New York Times*, 6 December 1997, B9.

34 Steven Pinker, 'Why They Kill Their Newborns', *New York Times Magazine* (2 November 1997), pp. 52–4.

35 Stephen Jay Gould, *The Mismeasure of Man* (New York, Norton, 1981).

36 Steven Rose, Richard Lewontin and Leo Kamin, *Not in Our Genes* (Harmondsworth, Penguin, 1984).

37 Tooby and Cosmides, in *The Adapted Mind*, p. 123, footnote 3.
38 Peter Singer, *A Darwinian Left: Politics Evolution and Co-operation* (London, Weidenfeld & Nicolson, 1999).
39 Matt Ridley, *The Origins of Virtue* (London, Viking, 1996), pp. 261–5.
40 Helena Cronin, 'It's Only Natural', *Red Pepper* (August 1997), p. 21.
41 Francis Fukuyama, *The End of Social Order* (London, Social Market Foundation, 1998).

8

Sewing Up the Mind:
The Claims of Evolutionary Psychology

Barbara Herrnstein Smith

Entitling a book *How the Mind Works*,[1] as Steven Pinker does, is already to presuppose a good bit of the answer: first, that 'the mind' is a discrete and readily distinguishable entity; second, that it operates mechanically; and, third, that its operations could be the object of a causal and presumptively scientific explanation. Each of these suppositions can be and has been questioned. In this chapter I focus on two aspects of the new – or maybe not-so-new – discipline of 'evolutionary psychology' as reflected in Pinker's book and its significant predecessor, *The Adapted Mind*:[2] first, the distinctly pre-emptive quality of the claims in that regard issued by its practitioners and promoters and, second, the peculiarly sutured image of the mind that emerges from the particular accounts they develop. Hence my title: 'sewing up' what has been called 'the mind', meaning amongst other things the processes of human cognition. Thus 'the mind' may be seen as a name given to a shifting set of heterogeneous phenomena and notions, ranging from observable patterns of behaviour and introspected experiences to the various faculties, processes and interior mechanisms that, at various times and in various informal and formal discourses (philosophical, ethical, legal, medical and so forth), have been posited or assumed to explain them. Accordingly, 'the mind' as such may not appear to be the sort of thing that cognitive scientists – as distinct from, say, intellectual historians – should seek to explain or ever could quite 'explain' at all.

Pinker is not unaware of such views, but dismisses them tartly early on (they amount, he writes, to saying that the mind is an obsolete fiction, like 'the Tooth Fairy')[3] and handles the questions they raise through facile definitions and *ad hoc* slides between 'mind' and 'brain' or between each of these and 'mental organs', 'neural circuitry', 'our thoughts and feelings', and much else besides. The conceptual problems thereby evaded, however, are significantly implicated in each of the two approaches that Pinker – and evolutionary psychology more generally – seeks to synthesise, namely, a strictly computational model of mind and a narrowly adaptationist account of human behaviour, and return, with other problems, to unsettle the ambitious claims made for the new discipline thus constituted.

Strategically equivocal on the relation between 'the mind' and 'the brain', Pinker neither quite identifies nor quite distinguishes between them. Thus, after declaring that 'the mind is what the brain does' (p. 24), he explains 'mental organs' as both complements of and analogues to 'bodily organs'. But if the mind is what the brain does, then it's not clear why we need mental organs at all. It would be the same as saying that birds have 'flying organs' as complements of and analogues to their wings. At other points, Pinker identifies mental organs with hard-wired (but as yet undetected) neural circuits, which would seem to make them more a matter of how the brain is constructed than of what it does. My concern here is not the equivocal usage as such, but its signalling of a more fundamental difficulty in the project of evolutionary psychology, namely, its effort – Quixotic, in my view – to provide a quasi-naturalised account of traditional Cartesian/rationalist conceptions of cognitive processes and, accordingly, its continuous need to finesse the mind/body problem.[4]

Comparable switches and slides recur in *The Adapted Mind*. For example, in their introductory essay to the volume, Jerome H. Barkow, Leda Cosmides and John Tooby write, 'The brain takes sensorily derived information from the environment as input, performs complex transformations on that information, and produces either data structures (i.e., representations) or behavior as output.'[5] Although reference is made here to what the *brain* does, the authors go on to stress the non-neurophysiological and apparently non-physical nature of the structures and mechanisms involved: '[Neuroscience] tells only how the physical components of the brain interact'; 'a cognitive . . . description [such as that offered by evolutionary psychology] characterizes the "programs" that govern its operation'.[6] The distinction is characteristic,[7] but cuts two ways: while it permits evolutionary psychology to claim autonomous status as a discipline, free to deduce mental 'programs' with minimal neurophysiological constraints, it also reveals the empirical limits or fundamentally non-empirical aspirations of the new discipline. Moreover, in this seemingly even-handed division of disciplinary labours ('physical components of the brain' or, as Pinker and others often put it, 'hardware' to neuroscience, 'governing "programs"' or 'software' to evolutionary psychology), the authors omit a significant project and achievement of contemporary neuroscience, namely its

ability to correlate observable and manipulatable physiological structures and events with observable and manipulatable behaviours and
reported experiences, and, crucially here, to account for such correla-
tions without the postulation of any intervening *mental* mechanisms.[8]
Indeed, in view of current work – theory as well as research – in
neuroscience, it appears that, whatever else may be said about the mind,
as a posited realm of interior, underlying causality it remains quite
hypothetical and arguably superfluous.

Describing the methodological centrepiece of the new discipline,
which he calls 'reverse engineering',[9] Pinker observes that, just as in the
case of an artifactual contrivance (his example is an olive pitter), so also
in the case of the mind: we can explain 'how it works' only by – and in –
identifying the purpose for which it was 'designed' and then, on the
basis of other assumptions, deducing how it must have been engineered
to achieve that purpose (pp. 21–2). Though psychologists, Pinker
informs his readers, have been unable even to approach this task
properly in the past (the history of the discipline being represented,
accordingly, as a series of bumblings and bungles), current practitioners
of evolutionary psychology, by virtue of their rigorous adherence to two
key ideas, have now almost completed it. The first is the idea that the
mind, like a computer, is an information-processing machine, the
purpose of which is to solve problems through rule-governed manipulations of symbols that represent objective features of the world. The
second is the idea that the mind, like the body, consists of multiple
individual organs or 'modules' engineered by natural selection to
maximise the reproductive fitness of our Upper Paleolithic ancestors
and reflecting that design more or less directly in their current
operations.

In his account and defence of the method, Pinker represents as a
sequence of logically linked productive inferences what could otherwise
be seen, both from his own examples and from the descriptions and
explanations of Tooby and Cosmides, as a process of self-enclosed
speculation directed by a set of mutually determining, mutually
validating assumptions, descriptions and hypotheses.[10]

For example, having first described vision as the brain's computational solution to 'the problem' of the mind's need to obtain 'accurate
knowledge of the world' from the 'shifting, impoverished' patterns of
light registered by the eyes, Pinker goes on to explain how evolutionary
psychologists deduce the existence of neural circuitry designed by
natural selection to perform computations to solve just that problem
(pp. 5–10). The putative 'problem' here, however, is an artefact of the
very computational model of mind that evolutionary psychology takes as

its initial premise. For it is only in accord with the realist/representationalist assumptions of that model that visual perception would be seen as defective informational input in the first place, or that the brain would be seen as requiring 'correct representations' of the world in order to direct behaviour appropriately. Other current models of cognition that describe visual (and other) perception without such ontological and epistemological presuppositions can also offer accounts of appropriate (that is, self-sustaining and effective) organism–environment interactions without the postulation or 'deduction' of extensive preloaded computational software.[11] Evolutionary psychology appears to be a theory that yields a method of analysis which generates the solution of problems created by the theory itself. Of course, it has some good company in those respects.

A crucial feature of the self-promotion of evolutionary psychology are claims to the effect that the programme and ideas it advocates are opposed only by such readily discountable adversaries as self-deluding theologians, sentimental humanists, political ideologues, social scientists clinging to 'archaic concepts of mind' (p. 57)[12] and, to be sure, 'postmodernists'.[13] Contrary, however, to the implications of such claims and lists, the founding ideas and assumptions of evolutionary psychology remain controversial within their originating disciplines and among cognitive scientists more generally.

Criticisms of the computational model of mind, itself developed largely by engineers, logicians and mathematicians, often reflect specifically biological considerations. Thus, biological-systems theorists and developmental psychologists note that prevailing versions of the model (other more or less radically modified versions continue to emerge) do not adequately reflect the peculiar structural and operational properties of living systems, including their characteristically global, self-regulating and, in important respects, nonlinear dynamics.[14] It has also been observed that, in embodied, mobile, socially embedded and verbal creatures such as ourselves, cognitive processes involve complex forms of social, verbal, perceptual and manipulative co-ordination as well as internal feedback mechanisms that, again, have no counterparts in the computational model as currently developed.[15] While Pinker attempts to discredit such criticisms by associating them with vitalism (life as 'a quivering, glowing, wondrous gel' [p. 22]), it is not a matter of honouring some ineffable distinction between organisms and physical systems but of understanding what kinds of physical systems organisms, including human beings, are.

Major alternative theoretical frameworks for explaining intelligent human behaviour and cognitive processes more generally give due weight to the relevant characteristics of living systems. Rather than a series of discrete problem-solving computations, the processes of cognition in human beings – as in organisms more generally – may be seen as the continuous modification of our structures and manners of functioning in the course of our interactions with an always, to some extent, changing environment. Proponents of such alternative ecological and/or dynamical models of cognition stress that human-like intelligent behaviour includes embodied skills that cannot be translated into symbolic manipulations or statement-like propositions, and that perception is an interactive activity, not a form of spectation or the passive reception of unidirectional environmental 'inputs'.[16] Computational models of mind can explain intelligent behaviour or cognition only to the extent that ongoing activities or processes can be decomposed into sets of context-free manipulations of discrete bits of 'information' in accord with supposedly prior and fixed 'rules'. From the perspective of these alternative accounts, this would be, at best, but a fraction of what might reasonably be understood as cognitive activities and processes.

Other criticisms of the computational model involve broader conceptual, including epistemological, considerations. For example, because human beings and their environments are mutually determining (that is, any creature's environmental niche – what it can interact with perceptually and behaviourally – depends on that creature's particular structure and modes of operating, and these are modified in turn by the creature's ongoing interactions with its environment), the objectivist, realist assumptions that the computational model inherits from rationalist philosophy of mind, in which the environment ('the world') is conceived in terms of fixed, autonomously determinate features, hobble its efforts to give a coherent account of the dynamics of cognition.

Dubious epistemological and indeed ontological presumptions are evident in Pinker's peremptory discussions of category formation, a question which is currently the subject of much debate among cognitive theorists.[17] Categories, he assures readers, are not 'arbitrary conventions that we learn along with other cultural accidents standardized in our language' – a view, certainly curious as stated, that he attributes to 'many anthropologists and philosophers' – but are, rather, 'forced' on us, via innate mental feature-detectors, by the way the world actually is: 'Mental boxes work because things come in clusters that fit the boxes' (p. 308). '[T]he dichotomy between "in nature" and "socially constructed",' Pinker observes at another point (the issue is social classifications, exemplified by 'the homosexual/heterosexual binary'), 'omits a third alternative: that some categories are products of a complex mind designed to mesh with what is in nature' (p. 57) – which is hardly a third alternative (though alternatives do, in fact, exist) or an escape from that dichotomy.

Pinker's related account of concept formation is exceedingly strained and comparably dubious. '[A] handful of concepts about places, paths, motions, agency and causation', he writes, 'underlie the literal or figurative meanings of tens of thousands of words and constructions' in all languages. The widespread recurrence of certain elementary ideas and relational concepts is generally granted; the question is how they come into being.[18] According to Pinker, they are innate and prewired, making up a universal conceptual lexicon – 'mentalese'[19] – the elements of which combine with 'input' from a local physical environment to form complex thoughts according to combinatorial rules that are also prewired. Connections between our ancestors' eyes, muscles and neural circuits, Pinker tells readers, were originally formed by natural selection 'for reasoning about rocks, sticks, space and force' but were subsequently 'severed', and 'references to the physical world were bleached out', so that the 'slots' (neural? mental?) thus left empty 'could be filled in with symbols for more abstract concepts' (pp. 355–6). An ingenious story. There is, however, no evidence for the existence of any such concept-specifying prewiring, much less for the occurrence of any such severing or bleaching-out processes – at least not in any of the relevant empirical fields, such as neurophysiology or palaeoanthropology. Indeed, like many other reverse-engineering deductions, the only ground for supposing the existence of such structures and processes is their explanatory necessity, given a prior commitment to nativist/ rationalist conceptions of human cognition and a determined neglect, dismissal or foreclosing of other avenues of research and theory.

Not surprisingly, analogies from the operations of computers figure largely in the computational model of mind. Reference to computers, Pinker maintains, 'demystifies' (but, significantly, also 'rehabilitates') common-sense beliefs about the mind by supplying 'hard-nosed' materialist, physicalist explanations of 'fuzzy' mentalistic concepts (pp. 78–9). To illustrate the point, he supplies a series of supposedly strict, precise translations: 'Beliefs are inscriptions in memory, desires are goal inscriptions, thinking is computation, perceptions are inscriptions triggered by sensors, trying is executing operations triggered by a goal' (p. 78). How much illumination is thereby provided, however, is questionable. 'Goal inscriptions' and 'inscriptions in memory' are, of course, no more material or physical (and no less reified) than 'desires' or 'beliefs', and none of these redescriptions is any more intrinsically 'hard-nosed' than the warehouse of other metaphors for cognitive processes drawn in the past from such technologies as agriculture, writing (the 'inscriptions' are obviously still there in force, though now recirculated through computers), hydraulics or cinematography. What Pinker exposes here, rather, is how a computational model of mind can give contemporary outfitting to traditional ideas and explanations without disturbing in the slightest their definitive – and, arguably, most problematic – features.

Although Pinker maintains that the computational model of mind is significantly validated by artificial intelligence, recent work in the field and in the related fields of artificial life and robotics tends to corroborate the other criticisms of the model outlined above. Most notably, researchers led by Rodney Brooks at the Artificial Intelligence Laboratory at MIT have engineered an impressive menagerie of intelligent-behaving automata that operate responsively and appropriately without previously installed rules, rule-governed manipulations or interior symbolic representations.[20] Pinker's omission of any mention of these achievements is conspicuous not only because Brooks is his institutional colleague but because the supposed limits of merely human robot-engineering provide a point of reference throughout *How the Mind Works*, highlighting by contrast the mind-boggling feats of mind-engineering that Pinker attributes to the super-human quasi-agent 'natural selection'.

Evolutionary theory is, of course, crucial to evolutionary psychology, but aspects of it are appropriated by Pinker and other evolutionary psychologists in dubious ways. For one thing, for its programme of 'reverse engineering' to make sense, every mental organ thereby deduced would have to be the end product of a series of genetic variants, each of which had conferred an adaptive advantage on members of the species possessing it. While evolutionary sequences of that kind can be reconstructed for many clearly identified perceptual capacities and behavioural traits (for example, the ability to recognise individual faces or the tendency towards nausea in the early weeks of pregnancy, for both of which Pinker cites persuasive adaptationist accounts), in the case of the complex behavioural traits and often totally putative mental capacities posited by evolutionary psychologists (for example, a supposed feature-detection device that guarantees the veridicality of the categories we form, or a supposed mental module for syntactic parsing), it is at best difficult to construct and often impossible even to hypothesise any such scenario that would be simultaneously genetically, neurophysiologically and ecologically plausible. Thus, as other evolutionary theorists point out, many of the explanatory accounts of evolutionary psychology, like those of its most immediate ancestor, sociobiology, require the assumption of prehistoric scenarios that range from the speculative to the unimaginable.

Related objections can be raised to the unremittingly purposive, rational idiom of evolutionary psychology: 'genes for', 'designed for', natural selection as the ever-ingenious 'engineer' of fitness optimising 'devices', and so forth. What is obscured here is the significance in both classic and contemporary Darwinian theory of historical contingency, that is, not only the randomness of genetic mutations or the possibility of rare accidents, such as comets falling to earth, but the everyday chanciness, particularity and dependency of all biological events, developmental as well as evolutionary. (See Stephen Jay Gould's chapter.) The problem is not the technically imprecise language but the crucial explanatory oversimplification. If, as Pinker complains, people who stress the interactive dynamics of individual development 'have a phobia about specifying what's innate' (p. 33), it is not – or not always and only, as he implies – because of political anxieties but because, especially with respect to the complex or highly mediated traits usually at issue (for example, human intelligence, language use or sexual preferences and behaviour), 'specifying what's innate' is precisely what cannot be done.[21]

Pinker has some serious phobias of his own, mostly focused on reminders of cultural and historical variability and the significance of individual experiences, social practices and normative institutions in shaping human behaviour. Invocations of such matters elicit his special scorn and he can hardly find terms derisive enough ('myths', 'Romantic nonsense', 'anthropological correctness', 'the common wisdom of Marxists, academic feminists and café intellectuals', and so forth [p. 431 *et passim*]) to characterise them. He is especially perturbed by the idea of *learning* (which he identifies, improperly and tendentiously, with mind-as-blank-slate empiricism)[22] and goes to great and often tortuous lengths to avoid acknowledging that our perceptual and behavioural tendencies – and, it may be presumed, neural circuits – are modified (strengthened, weakened, configured and reconfigured) throughout our lives by our interactions with our environments. For example, the discovery by developmental psychologists that stereoscopic vision is not present at birth, develops only gradually and is not always achieved does not challenge, for Pinker, the idea of its being a specific prewired capacity. Rather, it reveals for him the existence of a further supplementary mechanism, an alleged 'installation sequence' that programs the (pre)wiring to occur *after* birth. While such a process, he acknowledges, 'requires, at a critical juncture, the input of information that the genes can't predict', it is, he believes, 'a much better way of thinking about it' than 'the tiresome lesson that stereoscopic vision, like everything else, is a mixture of nature and nurture' (p. 238). It is hard to see, however, in what way that lesson – however crudely stated here by Pinker[23] – is thereby controverted or bettered. Indeed, the significance of, in effect, *postnatal* neural assembly is a key point of the ecological and developmental-systems models mentioned above, models that Pinker conflates (under the label 'the "interactionist" position') with simplistic nature-plus-nurture views and dismisses exasperatedly (and without specific citation) as 'useless', 'a colossal mistake', 'ideas so bad they are not even wrong', and so forth (pp. 32–3). Of course, he does not recognise his own duplication of that point here because, in the alternative models, the process(es) in question would be understood not as a fixed sequence (pre- plus post-natal) of discrete hardware installations for receiving informational inputs from the environment but as the ongoing environmentally responsive reconfiguration of systemic connections throughout an organism's lifetime. There is, moreover, a further disturbing possibility that Pinker might well wish to keep at bay. For, if the emergence of such an obviously advantageous competence as stereoscopic vision can be explained in this way (that is, as partly a matter of contingent epigenetic development or, in effect, learning), then it's not clear why many other cognitive capacities should not be, from cheater-detection to past-tense formation and, indeed, the whole array of handy human abilities that evolutionary psychology insists are specifically innately prewired. (See Patrick Bateson and Annette Karmiloff-Smith's chapters.)

The intellectual confinements that result from Pinker's assumptions of the innate and adaptive origins of human nature[24] are especially evident in his accounts of human social behaviour. Thus, while striking out at 'social construction[ism]' and 'fashionable "liberation" ideologies like those of Michel Foucault',[25] he evidently fails to grasp how some of the most elementary and ubiquitous instruments of human socialisation operate: for example, cultural representations such as images and lore, social sanctions such as shamings, and normative classifications sustained by institutions such as law or religion. He certainly underestimates their naturalising force. Thus, in support of his conviction that a biologically programmed repugnance rather than a culturally transmitted taboo 'explain[s] what keeps siblings apart', Pinker observes, without blinking, that '[b]rothers and sisters simply don't find each other appealing as sexual partners . . . the thought makes them acutely uncomfortable or fills them with disgust.' Biologically programmed? – maybe; 'simply'? – hardly.

The accounts of human social practices generated by evolutionary psychologists – from philanthropy and hypocrisy to courtship and spousal abuse – draw largely on concepts and related analyses developed by population geneticists and game theorists: 'inclusive fitness', 'reproductive investments', 'defections', 'payoffs', and so forth. In accord with such analyses, it is possible to calculate how certain behavioural choices (for example, co-operation rather than defection, submission versus flight or attack) would, under various hypothetical (and highly simplified) conditions, maximise the resources or enhance the reproductive fitness of the individuals pursuing them. Zoologists and behavioural ecologists use such calculations routinely to help explain a wide range of behaviours observed in other species. The idea that such analyses might also help explain various features of human social behaviour is not, in my view, mistaken in itself. What is mistaken, however, is the idea, played out on page after page of *How the Mind Works*, that there is something especially rigorous and scientific about transferring such calculations and analyses as rawly as possible from blackboard, barnyard and jungle – or presumed ancestral savannah – to contemporary human society.

'The human mating system', Pinker writes, 'is not like any other animal's. But that does not mean it escapes the laws governing mating systems, which have been documented in hundreds of species . . . For human sexuality to be "socially constructed" and independent of biology, as the popular academic view has it, not only must it have miraculously escaped these powerful [selection] pressures, but it must have withstood equally powerful pressures of another kind' (p. 467). I will not pause to deal with Pinker's distorted report of the idea of the social construction[26] or his related conflation of 'human sexuality' with 'gender roles' (see the passage quoted below). The more crucial and revealing point is his evident presumption that the social behaviours at issue have *not* withstood any such pressures or could have done so only 'miraculously'. The passage continues, 'If a person played out a socially constructed role, other people could shape the role to prosper at his or her expense.' Well . . . yes, that *could* happen. Indeed, it's a pretty good summary of the relations between the more powerful and less powerful members of the species throughout recorded human history (one thinks readily of feudalism, slavery or literal patriarchy). Pinker's implication here, however, is that such shaping did not and *could not* happen. His argument and supposedly clinching counter-example go as follows: 'Powerful men could brainwash the others to enjoy being celibate or cuckolded, leaving the women for them. Any willingness to accept socially-constructed gender roles would be selected out, and genes for resisting the roles would take over' (p. 467). Period. QED. Pinker evidently believes, and believes that he has just proved, that human beings have an innate, naturally selected resistance to the social construction of gender roles – or, indeed, given the logic of his argument, to the social construction of *any* roles.[27] Since, however, the proof depends on an obliteration of the contrary evidence of recorded human history (as distinct from the compliant evidence of unrecorded Upper Paleolithic history) and of the actual range of behaviours of contemporary human beings (as distinct from what 'any bartender or grandmother' will tell you),[28] all it demonstrates are the explanatory confinements of evolutionary psychology and the perceptual biases of at least one of its major practitioners.

Evolutionary psychology wants to call the mind-explaining game over and to declare itself and its team the winners. Contrary, however, to its portrayal of the scene, the remaining questions in cognitive science are not just technical, a matter of working out the details of a programme that all enlightened practitioners endorse. Quite the reverse: the field is exceptionally active at all levels – conceptual, empirical and methodological – and also both diverse and volatile, with new disciplinary configurations and domains of research opening up virtually continuously, and significant ideas and connections being developed on all sides.[29]

Like comparably ambitious projects from general equilibrium economics to structuralist linguistics, evolutionary psychology mistakes its own oversimplifications for the discovery of simplicity and its effacement of contingency, mediation and variability for the disclosure of universal laws. If the ongoing empirical programme in psychology is to develop scientific explanations of human behaviour and whatever other phenomena 'mind' is taken to include, it would not be compromised by acknowledging the cultural, historical and individual variables that Pinker and Barkow, Cosmides and Tooby dismiss. Similarly, due attention to the concepts, methods and findings of cultural anthropology, social history, sociology and related fields would not make evolutionary psychology any less 'hard-nosed' than it is now. On the contrary, such attention might introduce a degree of observational precision, documentary concreteness and conceptual subtlety (one thinks, for example, of Mary Douglas's classic analyses of pollution or Erving Goffman's fine-grained studies of social interaction) sorely missing at present from its accounts of human practices. Since the proponents of evolutionary psychology themselves appear captive to an unregenerate Two Cultures mentality, with its familiar intellectual provincialisms and disciplinary antagonisms, they are unlikely to see the advantage of such acknowledgements and attentions. As the exceedingly vital and protean field of cognitive science continues to define and configure itself, however, it may be hoped that new generations of scientists, scholars and theorists will energise its development in such directions.

Notes and References

1 Steven Pinker, *How the Mind Works* (New York, Norton, 1997; London, Allen Lane, 1997). The present essay is a revised and expanded version of an earlier review of the book. See Barbara Herrnstein Smith, 'Is It Really a Computer?', *Times Literary Supplement* (20 February 1998), pp. 3–4.

2 Jerome H. Barkow, Leda Cosmides and John Tooby (ed.), *The Adapted Mind: Evolutionary Psychology and the Generation of Culture* (Oxford, Oxford University Press, 1992).

3 Pinker, *How the Mind Works*, pp. 77–8. Subsequent page references appear in parentheses in the text.

4 For discussion of the problems encountered by other such efforts, see Barbara Herrnstein Smith, *Belief and Resistance: Dynamics of Contemporary Intellectual Controversy* (Cambridge, MA, Harvard University Press, 1997), pp. 141–4.

5 Barkow, Cosmides and Tooby, 'Evolutionary Psychology and Conceptual Integration', *The Adapted Mind*, p. 8.

6 Ibid.

7 The drawing of distinctions between, on the one hand, neurophysiological structures and processes and, on the other, the 'functional properties' of the mind is a mark of what is called, in cognitive science, *functionalism*: that is, the idea, developed largely by artificial-intelligence theorists and rationalist philosophers of mind, that the operations that define intelligence (or reasoning) do not depend on their embodiment in any particular physical medium, and could just as well be silicon chips in a computer as neurons in a living organism.

8 See, e.g., Gerald Edelman, *Bright Air, Brilliant Fire: On the Matter of the Mind* (New York, Basic Books, 1992), and Antonio Damasio, *Descartes' Error: Emotion, Reason, and the Human Brain* (New York, G. P. Putnam, 1994).

9 The method is referred to as 'evolutionary functional analysis' in the major theoretical chapter of *The Adapted Mind*, 'The Psychological Foundations of Culture', co-authored by John Tooby and Leda Cosmides.

10 Tooby and Cosmides's own explanations of the method are exceptionally angular and often misleading, with speculative reconstructions and hypotheses (e.g., of Upper Palaeolithic environmental conditions or specific putative mental mechanisms) referred to as 'descriptions' and what should, accordingly, be modal terms (*if, would, were to*, etc.) replaced by quasi-observational terms ('when', 'did', 'does', etc.). The final component of the method, which the authors refer to as a 'performance evaluation', is described as if it were an empirical test: '[It is] important . . . to see whether

the proposed mechanism produces the behaviors one actually observes from [sic] real organisms in modern conditions. If it does, this suggests the research is converging on a correct description of the design of the mechanisms involved' (*The Adapted Mind*, p. 74). But, of course, one cannot 'see', in the sense of observe, whether a hypothetical mental mechanism (e.g., a putative language-acquisition device or cheater-detection module) produces some actual behaviour; one can only assert that it does so on the basis of the assumptions, observations and speculations that led one to posit it in the first place. Indeed, as quasi-validation procedure, this component of the method appears to be a virtual prescription for self-affirming circularity.

11 See, e.g., Humberto R. Maturana and Francisco J. Varela, *Autopoiesis and Cognition: The Realization of the Living* (Boston, D. Reidel, 1980) and *The Tree of Knowledge: The Biological Roots of Human Understanding* (Boston and London, Shambala, 1988); Esther Thelen and Linda B. Smith, *A Dynamic Systems Approach to the Development of Cognition and Action* (Cambridge, MA, MIT Press, 1994).

12 For sceptical discussion of what Pinker, following Barkow, Cosmides and Tooby, alleges to be a 'Standard Social Science Model', see Hilary Rose's chapter.

13 Pinker attributes residual objections to the programme of evolutionary psychology to 'the takeover of humanities departments by the doctrines of postmodernism, poststructuralism, and deconstruction, according to which objectivity is impossible, meaning is self-contradictory, and reality is socially constructed' (p. 57). There are no citations for these ridiculous claims, ignorant conflations and glib characterisations. For comparable claims, conflations and dismissals in connection with a related and comparably pre-emptive programme, see E. O. Wilson, *Consilience: The Unity of Knowledge* (New York, Knopf, 1988), pp. 40–4.

14 See, e.g., Susan Oyama, *The Ontogeny of Information: Developmental Systems and Evolution* (Cambridge, Cambridge University Press, 1985); Patrick Bateson, 'Biological Approaches to the Study of Behavioral Development', *International Journal of Behavioral Development*, 10: 1 (1987), pp. 1–22; Thelen and Smith, *A Dynamic Systems Approach to the Development of Cognition and Action*; Tim van Gelder, 'What Might Cognition Be, If Not Computation?', *Journal of Philosophy*, XCI: 7 (July 1995), pp. 345–81; and Tim van Gelder and Robert F. Port, 'It's About Time: An Overview of the Dynamical Approach to Cognition', in van Gelder and Port (ed.), *Mind as Motion: Explorations in the Dynamics of Cognition* (Cambridge, MA, MIT Press, 1995).

15 See, e.g., Lucy Suchman, *Plans and Situated Actions* (Cambridge, Cambridge University Press, 1987); Edwin Hutchins, *Cognition in the Wild* (Cambridge, MA, MIT Press, 1996); Tim Ingold, 'Technology, Language, Intelligence: A Reconsideration of Basic Concepts', in K. R. Gibson and T. Ingold, *Tools, Language and Cognition in Human Evolution* (Cambridge, Cambridge University Press, 1993), pp. 449–72; and Horst Hendriks-

Jansen, *Catching Ourselves in the Act: Situated Activity, Interactive Emergence, Evolution and Human Thought* (Cambridge, MA, MIT Press, 1996).

16 See Francisco J. Varela, Evan Thompson and Eleanor Rosch, *The Embodied Mind: Cognitive Science and Human Experience* (Cambridge, MA, MIT Press, 1991); A. Blake and A. Yuille (ed.), *Active Vision* (Cambridge, MA, MIT Press, 1992); Maturana and Varela, *Autopoiesis*; Oyama, *The Ontogeny of Information*; and Thelen and Smith, *A Dynamic Systems Approach*.

17 See especially George Lakoff, *Women, Fire, and Dangerous Things: What Categories Reveal About the Human Mind* (Chicago, Chicago University Press, 1987). For recent discussion and an important contribution to the debate, see Paul E. Griffiths, *What Emotions Really Are: The Problem of Psychological Categories* (Chicago, University of Chicago Press, 1997).

18 Although Pinker cites George Lakoff and Mark Johnson, *Metaphors We Live By* (Chicago, University of Chicago Press, 1980), to illustrate the existence of such concepts, he omits mention of Lakoff and Johnson's pointedly anti-innatist account of them as arising from recurrent bodily experiences.

19 The term 'mentalese' and much else in Pinker's account of concept formation is adopted from the work of philosopher/cognitive theorist Jerry A. Fodor. For sceptical assessments of Fodor's views by other cognitive theorists and philosophers, see Andy Clark, *Associative Engines: Connectionism, Concepts and Representational Change* (Cambridge, MA, MIT Press, 1993); Hendriks-Jansen, *Catching Ourselves in the Act*; and Stephen Schiffer, 'Lessons in Mentalese: Fodor's Theory of Mind and its Problems' (review of Fodor, *Concepts: Where Cognitive Science Went Wrong*, *Times Literary Supplement* [26 June 1998], pp. 14–15).

20 In a series of influential papers published over the past decade, Brooks and his associates examine and criticise each of these key features of the standard computational model and indicate why they discard them. See, e.g., Rodney Brooks, 'Intelligence without Representation', *Artificial Intelligence*, 47 (1991), pp. 139–59.

21 For relevant critiques of innatist and/or genocentric accounts of such traits, see Terrence W. Deacon, *The Symbolic Species: The Co-evolution of Language and the Brain* (New York and London, Norton, 1997), pp. 334–40; Philip Lieberman, *Eve Spoke: Human Language and Human Evolution* (New York, Norton, 1998), pp. 125–32; and Susan Oyama, *Evolution's Eye* (Durham, NC, Duke University Press, forthcoming).

22 According to Pinker, what is commonly referred to as 'learning' is actually the product of specific prewired mechanisms for acquiring particular behaviours (e.g., language use). 'It [learning] is made possible by innate machinery designed to do the learning' (*How the Mind Works*, p. 33).

23 'Mixture', which improperly suggests simple addition, should be *interplay;* 'nature and nurture', a relatively uninformed cliché, might be better – more precisely and less 'tiresome[ly]' – framed as *genetic and epigenetic processes*.

24 These are detailed in Pinker's earlier book, *The Language Instinct: How the Mind Creates Language* (New York, William Morrow, 1994).

25 Pinker's knowledge of these matters appears, from his garbling of them, to be second-hand – derived mainly, his citations suggest, from hostile journalistic reports.

26 Pinker is either being obtuse here or taking advantage of the multiple meanings of 'sexuality' and 'biology' to foist a manifestly absurd claim on some of his critics. Although the idea of 'social construction' can be invoked and applied crudely (as can any other idea, including 'natural selection'), the point usually made by those citing it in these connections is not that people's physiological traits, erotic feelings or carnal activities ('human sexuality' in any of these senses) are independent of the evolutionary history of the species or of those people's individual genomes ('biology' in either of those senses) but, rather, that conventional divisions, associations and normative attributions of various traits, feelings, activities and roles – especially in accord with simple male/female or homosexual/heterosexual dichotomies – are not simple products or direct 'expressions' of either that history or those genomes.

27 This supposed refutation of the idea that human sexuality (in whatever sense) could be socially shaped cleans the slate for Pinker's alternative account of its definitive determination (in presumably all senses) by the fitness-enhancing strategies of our mammalian and Upper Palaeolithic ancestors. The immediately ensuing paragraph begins, 'What kind of animal is *Homo sapiens*? We are mammals, so a woman's minimal parental investment is much larger than a man's. She contributes nine months of pregnancy and (in a natural [sic] environment) two to four years of nursing. He contributes a few minutes of sex and a teaspoon of semen . . .' After further details of this sort, Pinker observes, 'These facts of life have never changed . . . These conditions persisted through ninety-nine per cent of our evolutionary history and have shaped our sexuality . . . A part of the male mind, then, should want a variety of sexual partners for the sheer sake of having a variety of sexual partners . . . Any bartender or grandmother would say . . .' and so forth (*How the Mind Works*, pp. 468–9).

28 Pinker adds to their testimony (see note above) the results of a statistical survey of the dating and mating preferences of contemporary college men and women.

29 For an instructive and discriminating recent survey, see Hendriks-Jansen, *Catching Ourselves in the Act.*

9

Why Babies' Brains Are Not
Swiss Army Knives

Annette Karmiloff-Smith

Evolutionary Psychology and Nativism: A Marriage Made in Heaven?

Evolutionary psychologists and nativists speak the same language – an inappropriate one, in my view. Evolutionary psychology claims that the millions of years of evolution have resulted in our brains becoming increasingly complex and 'pre-specified'. Nativists, who are committed to the view that many aspects of human behaviour are genetically determined, endorse this view.[1,2] Both assert, therefore, that the human brain is innately pre-specified not only for low-level perceptual processes such as vision, but also for such higher-level cognitive functions as the capacity to acquire and use language and number, and to recognise faces. Furthermore, they argue that each mental domain functions independently of the others. This domain specificity is often referred to as 'the modular view of the mind'.[3]

I agree that human evolution has indeed led to increasingly complex forms of behaviour. However, these behaviours are not simply triggered from genetically determined mechanisms. Rather they are the outcome of the gradual formation of internal representations during the lengthy process of ontogenetic development. It is in the part of the brain called the neocortex (evolutionarily the most recent part of the brain) that higher cognitive functions like language, number and faces are processed. Rather than genetically prespecified complex, domain-specific representations, evolution may have generated an ever-increasing range of different learning mechanisms in order to ensure adaptive outcomes. This would suggest that during postnatal development the brain has a substantial capacity to learn and actively structure its own circuits while engaged in processing different types of environmental input. Of course, the infant brain has some pre-specified anatomical structure, such as the six layers of neuronal cells found in each region of the brain cortex. This is the macro-structure of the brain. However, the brain's micro-structure, that is the vast number of connections between cells and between different regions, is constructed epigenetically during the lengthy period of postnatal brain development.[4] (See Steven Rose's chapter.)

Although evolution may have resulted in the brain possessing a number of different *domain-relevant* learning mechanisms, such as the ability to process both sequential and holistic, simultaneous input, these mechanisms only become *domain specific*, i.e., applied specifically to, say, a higher cognitive function like language, as a function of actually working in that particular domain. Thus I argue, by contrast with the nativists, that it is the process of development (that is ontogeny itself) that progressively creates higher cognitive functions. In fact, some complex, higher-level cognitive outcomes may not be possible without this gradual process.[4]

Unfair to Nativism?

Critics of nativism are often met with the response that no nativist ever makes such fundamentalist claims and that their critics are first constructing and then demolishing a myth.[5] Here are a representative set of quotations from influential figures in the field:

It is uncontroversial that the development [of Universal Grammar] is essentially guided by a biological, genetically determined program. . . . Experience-dependent variation in biological structures or processes . . . is an exception . . . and is called 'plasticity'.[6]

We argue that human reasoning is guided by a collection of innate domain-specific systems of knowledge.[7]

syntactic knowledge is in large part innately specified.[8]

[the human mind is] equipped with a body of genetically determined information specific to Universal Grammar.[9]

language instinct . . . a Nature-given gift to speak complex, fluent grammatical language . . . This corresponds to the child innately expecting there to be, say, suffixes for persons, numbers, tenses and aspect, as well as possible irregular words, but not knowing exactly which combinations, suffixes or irregulars are found in the particular language.[10]

The mind is likely to contain blueprints for grammatical rules ... and a special set of genes that help wire it in place.[10]

If language, the quintessential higher cognitive process, is an instinct, maybe the rest of cognition is a bunch of instincts too – complex circuits designed by natural selection, each dedicated to solving a particular family of problems.[10]

Clearly some influential nativists believe in innately specified, domain-specific representations. Critics are not fabricating myths; the nativists are propagating them. Those like me who work on child development see plasticity during brain growth as the rule, not simply as an exception or response to injury[4].

Nativist claims are very much in line with those of evolutionary psychology, as the last citation, from Steven Pinker, illustrates. Evolutionary psychologists compare the newborn brain to a Swiss army knife, crammed with independently functioning tools, each designed for a specific problem that faced our hunter-gatherer ancestors.[2] Even if we set to one side the problem of knowing just what the problems were that our ancestors faced, and therefore what were and are the tools required, the claim is by no means unquestionable. Is it not equally plausible to hypothesise that, unlike the gross macro-structure of the brain, cortical micro-circuitry is not innately specified by evolution but is progressively constructed by the postnatal experience of processing different kinds of input?

Can Findings from Neuropsychology Address the Claims of Evolutionary Psychology?

One way of relating brain functions to structures is by studying the effects of brain damage, caused by stroke or accident. Often these result in very specific loss of particular functions. Sometimes damage to one brain region produces one type of loss whilst sparing another, and by contrast damage to another nearby region will have the opposite effect. This is called a double dissociation. For instance, in one case grammar may be impaired whilst vocabulary remains intact and, in another, exactly the opposite pattern may occur; a person will have great difficulty in accessing words, but grammatical structures remain intact. On witnessing a man leap out of a window, the first patient might only be capable of uttering something like: 'man jump window get hurt', whereas the second might produce grammatical markers like 'the' and '-ing', but experience huge difficulties in finding the right words: 'the uhm ... human uhm ... male ... what's the word for that ... is throwing ... no ... junning ... no ... out of the uhm ... thing uhm to open ... to look through uhm ... door, no, willow ... no ...' In other words, one patient can still use vocabulary but not grammar, and the other still has access to English grammar but cannot find the nouns and verbs necessary to tell the story. Such double dissociations have been used to make claims about the modular structure of the brain.

Other double dissociations occur, for instance, in the case of face processing. One patient may be severely impaired in recognising familiar faces but have no problems in recognising objects of all kinds. Again, a double dissociation occurs when another patient is found who shows the opposite problem: intact face processing but an inability to recognise other familiar visual stimuli like buildings or objects in a kitchen. It is the existence of these double dissociations that has been used by theorists to make claims about the modular structure of the brain, i.e., that it is made up of isolated, specialised systems called modules that can be differentially impaired or spared.

Although it is rare to find such pure cases, I have no problem with the logic of the argument as it pertains to the adult brain. Clearly, when an already structured, normal brain becomes damaged in adulthood, the damage may well impair specialised areas of processing. But this does not necessarily mean that the brain started out with these specialised circuits already in place. It could be that specialisation builds up gradually and is actually the *product* of child development, not its starting point. So even if modules were identified in damaged adult brains, this in no way entails that they were prespecified by evolution in the newborn brain. The modules could have emerged from a process of gradual specialisation or 'modularisation' during earlier ontogeny, prior to the damage in adulthood.[11] In other words, domain-specific outcomes do not necessarily entail domain-specific origins. So the study of double dissociations in damaged adult brains cannot be used to directly address the questions posed by evolutionary psychology. However, it remains possible that insights relevant to evolutionary claims could be drawn from dissociations found in developmental disorders in childhood.

Can the Findings from Studies of Abnormal Phenotypes Address the Claims of Evolutionary Psychology?

How can the study of atypical development address the evolutionary psychology question? I believe it can, but not in its currently popular form of approach. This is because much of the research on developmental disorders is studied within a static neuropsychological model more appropriate to adults. Paradoxically, developmental disorders are rarely studied developmentally.[12, 13] When researchers find a condition in which there seems to be an isolated impairment, such as in the case of specific impairments in language which can coexist with normal intelligence, they tend to conclude that this is the proof that there must be a genetically specified module for grammar.[14] A further (albeit implicit) assumption is that, despite the genetic defect and the purported missing or damaged module, the rest of the brain has developed normally. This ignores the dynamics of brain development and the way in which different parts of the brain interact as they grow. In other words, the abnormal brain is not a normal brain with parts intact and parts impaired. It is a brain that develops differently throughout embryogenesis and postnatal brain growth.

Nonetheless, many researchers focus on developmental disorders specifically to make claims to bolster the nativist stance.[5,14] There are indeed cases that at first blush would seem to fit this picture. In individuals with mild autism, tasks involving what is known within developmental psychology as 'theory of mind' – that is, the ability to impute intentional states to others – are seriously impaired. By contrast other aspects of the cognitive system appear to be relatively preserved.[15] Furthermore, in certain subgroups of children with the condition known as Specific Language Impairment, serious grammatical impairments seem to coexist with normal functioning in all other domains.[14] A further example comes from Williams syndrome (WS), a genetically caused developmental brain disorder.[12] People with WS show good levels of proficiency in face processing, language and social interaction, in sharp contrast to their severe impairments in spatial cognition, number and problem solving. The dissociations between impaired and seemingly preserved domains in WS are typical of those also described for brain-damaged adults. But do such findings necessarily imply domain-specific beginnings in which some modules are preserved and others impaired?

Williams Syndrome: An Example of Modular Beginnings?

Most studies of atypical development focus on areas of impairment, e.g., reading in dyslexia, social interaction in autism, or grammar in Specific Language Impairment. However, to use dissociations in developmental disorders to support or challenge the nativist claim, it is necessary to be certain that the purported *non*-impaired domains are indeed intact. To test this, I and my colleagues made a series of studies of infants, children and adults with WS, focusing on two of their particular areas of proficiency: language and face processing. These domains, in which people with WS score well (often even in the normal range for their chronological age), have frequently been used to make claims about the dissociation of general intelligence from language and the dissociation of visual-spatial processing from face recognition. The claims are then generalised to argue in favour of innately specified, specialised brain circuits for face processing and language.[10,16] Behind such claims is the notion that the missing genes in WS (caused by a microdeletion on the long arm of chromosome 7 [12] have no effect whatsoever on some domains and serious effects on the others. It is as if gene expression were a one-to-one mapping between gene and behavioural outcome.

Our aim in focusing on areas of proficiency was to discover whether the cognitive processes underlying proficient, overt behaviour are the same in WS as in normal development. If they are, then the evolutionary claim might indeed be correct. However, if we could demonstrate that people with WS solve language and face-processing tasks by processes that are different from those used in normal development, then their relatively intact *behaviour* cannot be used to make evolutionary claims about the prespecification of domain-specific modules.

In the event, our study showed that children and adults with WS reach their behavioural mastery via different processes from normal controls. For example, a number of earlier studies had demonstrated that in tasks requiring face processing and memory,[17,18] WS scores are in the normal range.[16] When we analysed WS face-processing capacities in more detail, distinguishing between recognition of face identity, facial emotion, eye-gaze direction and lip reading,[20] we found that WS participants did almost as well as the normal controls on some of the items, but extremely poorly on others.[20,21] The controls did equally well on both. When we further analysed the successes and failures of the WS group, we found that the items on which they achieved good scores were those that could be solved via a strategy in which the individual components could be analysed by a feature-by-feature strategy. Those on which they failed required configural or holistic processes. The normal controls tended to use configural processes for both sets of items.

Our findings have since been confirmed by asking subjects to identify right-way-up and upside-down faces. Normal controls display much greater difficulty with upside-down (inverted) faces, whereas people with WS tend to perform almost as well on both.[22] It is assumed that configural processing cannot be used with inverted faces. This explains the difficulty experienced by the normal controls. By contrast, if the subject is highly practised at feature-by-feature processing, as is the presumed case of people with WS, then upright and inverted faces both pose similar problems. Furthermore, the brain processes of people with WS, when identifying faces, show far less right hemisphere specialisation than normal controls and a different pattern of electrical activity. (23)

The excellent scores on the face recognition and memory tasks highlight the fact that even though *behaviourally* people with WS seem to be a prime example of the sparing of a face-processing module fashioned by evolution, and on which their retardation in spatial cognition has no effect, it turns out that they are in fact successfully solving the face-processing tasks via a different strategy. Thus their successful behaviour cannot be invoked to argue in favour of a preserved, innately specified face-processing module, even if such a module might normally exist. Empirical research on WS gives no support to the evolutionary psychology claim.

Much of the above argument about face processing holds also for the proficient levels of language displayed by people with WS. People with WS can learn large and rather erudite vocabularies and use quite complex grammatical structures.[16] This is very atypical of any other group with IQs in the 50s to 60s range. However, when we probed WS subjects' language ability in more detail we found that the processes by which children and adults with WS learn new words do not obey the same lexical constraints as do normal children's.[24] For instance, normal children expect new words to refer to whole objects unless they already know the name of the object. People with WS, by contrast, take a new word to refer just as readily to a part of an object, say, its handle. As in the case of face processing, they show an atypical tendency to focus on parts. Furthermore, when processing a number of grammatical structures (relative clauses, sentences with transitive and intransitive verbs), individuals with WS are not sensitive to all the same grammatical violations as normal controls.[25] They even have difficulty simply repeating a short sentence if its structure is complex —

for example, an embedded relative clause such as 'the boy the dog chases is big'.[26]

Repeatedly we find that once we probe more deeply the superficial proficiency of WS language we find subtle impairments and different processes from the normal controls. Once again, as with the case of face processing, these findings are supported by brain-imaging studies which have shown different patterns in WS adults compared to normal controls at all ages.[27] All of the above suggests that people with WS do not simply call on an intact language module fashioned by evolution and innately specified. Rather, people with WS appear to follow a deviant developmental pathway in their language acquisition.

Generalisations can be dangerous, though. Not all behavioural syndromes arise through following a deviant developmental pathway; others may be the consequence of developmental delay instead. Thus cognitive researchers need to exercise caution in drawing conclusions directly from behaviour, for it cannot be mapped directly on to underlying cognitive processes. Hence behavioural syndromes cannot be used directly to support the claims of nativists and evolutionary psychologists. It is unlikely that atypical development will be explicable in terms of preserved or damaged whole modules that have been genetically prespecified. Rather, subtle differences in developmental timing, numbers of neurons and their connections, transmitter types and neural efficiency are likely to be the causes of dissociations in developmental outcomes.[4,12] Such subtle initial differences may have a huge but very indirect impact on the resulting phenotype, ultimately giving rise to domain-specific impairments after the process of postnatal brain development. It is therefore crucial to go back to the origins of development in infancy.

Can the Findings from Normal and Atypical Infant Development Address the Evolutionary Question?

Nativists and evolutionary psychologists tend to make use of precocious abilities in infants to support their claims. But is the fact that children can perform certain tasks during infancy proof that the ability must be innately specified? Studies of normal infant development in fact show that both specialisation and localisation are very *gradual*.[28,29] Research on six-month- and twelve-month-old infants' face-processing capacities provide a good illustration. Brain-imaging experiments with two groups of infants shown a series of upright and inverted faces revealed that the younger group (six-month-olds) process faces using several areas of the brain and across both hemispheres. By twelve months of age, in contrast, face-processing circuits have become far more specialised and more localised in the right hemisphere.[28] In other words, it isn't until about twelve months of age that infants' face-processing resembles that of an adult. One could claim that the face-processing module was only 'turned on' at twelve months, i.e., that it is under maturational control. But surely a more parsimonious and more likely explanation is that by twelve months the infant has had sufficient experience of faces to cause the microcircuits in the neocortex to become progressively specialised and localised. Then, later in life when an adult suffers brain damage, there could well be a clear-cut dissociation with other visual-spatial processes because the later damage may knock out a localised circuit which had gradually become specialised during childhood.

Similar results are obtained with respect to gradual specialisation and localisation of language in normal development.[29] Brain-imaging studies of normal children have shown that until they are about six years of age, grammar is processed bilaterally by both hemispheres. It is only after several years that the left hemisphere becomes more specialised and localised for grammar (in right-handed individuals). That is why, in the rare cases in which the left hemisphere has to be removed to treat epilepsy, the development of language in the single, remaining right hemisphere has a far better prognosis if the operation is carried out early in the developmental process.[30] It is relevant that children with WS do not manifest normal progressive hemispheric specialisation and localisation. Once again, this points to gradual specialisation as a function of the processing of linguistic input during ontogenesis, not to rigid prespecification via evolution. It is thus essential to consider the developmental dimension, rather than simply the end state.

We have recently been studying the infant origins of atypical development in WS and in Down's syndrome. Using our knowledge of the different end states of these two abnormal phenotypes, we have been attempting to relate the end states to the starting states. Is it possible simply to generalise from the end state to the initial state? In my view, this is dangerous. People with WS end up with linguistic behaviour that is far superior to their spatial cognition. People with Down's show the opposite pattern. People with either of these syndromes display serious impairments in numerical reasoning. From these end states, one might expect WS infants to show a superior pattern to Down's infants in language, and children with either syndrome to display similar impairments in number. Our preliminary results suggest that this is not the case.[31] On an infant task testing vocabulary development where one might expect the WS infants to do well, in fact WS and Down's infants are equally delayed and look identical to their younger mental age-matched controls. By contrast, on a number task in infancy WS children are well in advance of those with Down's and behave like their chronological age-matched controls, despite both phenotypes being equally impaired in numerical reasoning in the end state. What changes, then, is the subsequent process of learning in each domain that results in the differing cognitive capacities observed in adults suffering from either of the two conditions. This once again stresses the importance of taking a thoroughly developmental stance and exercising caution when seeking to use either normal or atypical infant behaviour to support evolutionary psychology and nativist claims.

Evolution Versus Ontogenetic Outcomes?

Two alternative developmental models capture the difference between nativist and evolutionary psychological claims and those developed in this chapter: mosaic and regulatory.[4] Nativists would argue for mosaic development. It is under tight genetic control, fast, involves the independent development of different parts of the system and is fine under optimal conditions. However, more or less everything must be specified in advance and there are upper bounds on complexity. Some species do indeed follow mosaic development and some parts of all development are mosaic in nature, that is, their epigenesis (their genetically determined development) is indeed deterministic. This may be true of some parts of the human brain, but it does not hold for the development of cortical functions. Regulatory development is far more common and is certainly more typical of the probabilistic epigenesis of the development of the human cortex – the seat of higher cognitive functions. Regulatory development is under broad genetic control; it is slow and progressive, and numerous parts of the system develop interdependently. Like mosaic development, it is vulnerable under non-optimal conditions, but, most importantly, it is flexible in the face of change. The prespecification of regulatory development is minimal, but in processing the environmental input its resulting complexity is much greater than mosaic development. Regulatory development has far fewer upper bounds on complexity.

In this chapter, I have attempted to make the case that, despite the claims of nativism and evolutionary psychology, regulatory development is far more typical of what happens during cortical brain growth in human infants and children. Differential timing of developmental events and multiple levels of interaction both serve progressively to *construct* complexity rather than building it in advance. Evolution has helped to guarantee human survival by raising the upper limits on complexity and avoiding too much prespecification of higher cognitive functions. The answer to the false dichotomy cannot be evolution *or* ontogeny, as clearly development demands both. But nativists and evolutionary psychologists with their one-sided approach need to give far more consideration to the importance of the gradual process of ontogeny as the child interacts with the wealth of environmental inputs. This does not constitute a return to the discredited psychological tradition of behaviourism or the notion that brains possess a general problem solver for all domains. A multitude of different learning mechanisms which may have emerged from evolution might, during ontogeny, each discover inputs from the environment that are more or less suited to their form of processing. Gradually, with development and with trying to process different kinds of input, each mechanism would become progressively more domain specific.

Evolutionary psychology's Swiss army knife view of the brain is inappropriate for understanding higher cognitive functions, particularly for children. Indeed if we want to hang on to the Swiss army knife metaphor, then we need to shift focus from the end product, the clever little knife sold in the shops, to the account of its development. Thus, we should explore how each of the specialised parts of the army knife develops over time from a series of simpler tools. That is, the knife's ontogenesis.

References

1 Steven Pinker, *How the Mind Works* (New York, Norton, 1997).
2 Leda Cosmides and John Tooby, 'Beyond Intuition and Instinct Blindness: Toward an Evolutionary Rigorous Cognitive Science', *Cognition*, 50 (1994), pp. 41–77.
3 Jerry Fodor, *The Modularity of Mind* (Cambridge, MA, MIT Press, 1983).
4 Jeffrey Elman et al., *Rethinking Innateness: A Developmental Perspective on Connectionism* (Cambridge, MA, MIT Press, 1996).
5 Alan Leslie, 'Inhibitory Processes in Understanding False Belief', talk presented at the Conference on Modular and Constructivist Perspectives on Normal and Atypical Development, University College London (September 1998).
6 Kenneth Wexler, 'The Development of Inflection in a Biologically Based Theory of Language Acquisition', in M. Rice (ed.), *Towards a Genetics of Language* (Mahway, NJ, Lawrence Erlbaum, 1996), pp. 113–44.
7 Susan Carey and Elizabeth Spelke, 'Domain-Specific Knowledge and

Conceptual Change, in L. A. Hirschfeld and S. A. Gelman (ed.), *Mapping the Mind: Domain Specificity in Cognition and Culture* (Cambridge, Cambridge University Press, 1994), pp. 169–200.

8 Stephen Crain, 'Language Acquisition in the Absence of Experience', *Behavioral and Brain Sciences*, 14 (1991), pp. 591–611.

9 Neil Smith and Ianthi Tsimpli, *The Mind of a Savant: Language Learning and Modularity* (Oxford, Blackwell, 1995).

10 Steven Pinker, *The Language Instinct* (New York, William Morrow, 1994).

11 Annette Karmiloff-Smith, *Beyond Modularity: A Developmental Perspective on Cognitive Science* (Cambridge, MA, MIT Press, 1992).

12 Annette Karmiloff-Smith, 'Development Itself Is the Key to Understanding Developmental Disorders', *Trends in Cognitive Sciences*, 2, 10 (1998), pp. 389–98.

13 Dorothy Bishop, *Uncommon Understanding* (Hove, Psychology Press/Erlbaum, 1997).

14 Myrna Gopnik, 'Feature-Blind Grammar and Dysphasia', *Nature*, 344 (1990), p. 715.

15 Alan Leslie, 'Pretense, Autism and the "Theory of Mind" module', *Current Directions in Psychological Science*, 1 (1992), pp. 18–21.

16 Ursula Bellugi et al., 'Williams Syndrome: An Unusual Neuropsychological Profile', in S. H. Broman and J. Grafman (ed.), *Atypical Cognitive Deficits in Developmental Disorders: Implications for Brain Function* (Hillsdale, NJ, Erlbaum, 1994), pp. 23–56.

17 Arthur Benton, et al., *Benton Test of Facial Recognition* (New York, Oxford University Press, 1983).

18 Barbara Wilson et al., *The Rivermead Behavioural Memory Test* (Fareham, Thames Valley Test Company, 1985).

19 Orlee Udwin and William Yule, 'A Cognitive and Behavioural Phenotype in Williams Syndrome', *Journal of Clinical and Experimental Neuropsychology*, 13 (1991), pp. 232–44.

20 Robin Campbell et al., *Testing Face Processing Skills in Children* (Stirling University, 1995).

21 Frances Djabri, *Face Processing in Williams Syndrome: A Further Case of Within-Domain Dissociations* (unpublished undergraduate student project, University College London, 1995).

22 Annette Karmiloff-Smith, 'Crucial Differences Between Developmental Cognitive Neuroscience and Adult Neuropsychology, *Developmental Neuropsychology*, 13 (1997), pp. 513–24.

23 Christine Deruelle et al., 'Configural and Local Processing of Faces in Children with Williams Syndrome', *Brain and Cognition*, 41 (1999), pp. 276–98.

24 Helen Neville et al., 'Effects of Altered Auditory Sensitivity and Age of Language Acquisition on the Development of Language-Relevant Neural Systems: Preliminary Studies of Williams Syndrome', in S. Broman and J. Grafman (ed.), *Cognitive Deficits in Developmental Disorders: Implications for Brain Function* (Hillsdale, NJ, Erlbaum, 1993), pp. 67–83.

25 Tassos Stevens and Annette Karmiloff-Smith, 'Word Learning in a Special Population: Do Individuals with Williams Syndrome Obey Lexical Constraints?', *Journal of Child Language*, 24 (1997), pp. 737–65.

26 Annette Karmiloff-Smith et al., 'Linguistic Dissociations in Williams Syndrome: Evaluating Receptive Syntax in On-Line and Off-Line Tasks', *Neuropsychologia*, 36 (1998), pp. 343–51.

27 Annette Karmiloff-Smith et al., 'Syntax in Williams Syndrome: The Case of the Relative Clause' (1999, submitted).

28 Helen Neville et al., 'Effects of Altered Auditory Sensitivity and Age of Language Acquisition on the Development of Language-Relevant Neural Systems: Preliminary Studies of Williams Syndrome', in S. Broman and J. Grafman (ed.), *Cognitive Deficits in Developmental Disorders: Implications for Brain Function* (Hillsdale, NJ, Erlbaum, 1993), pp. 67–83.

29 Mark Johnson, *Developmental Cognitive Neuroscience: An Introduction* (Oxford, Blackwell, 1997).

30 Debra Mills et al., 'Variability in Cerebral Organization During Primary Language Acquisition', in G. Dawson, K. Fischer et al. (ed.), *Human Behavior and the Developing Brain* (New York, Guilford, 1994), pp. 427–55.

31 F. Vargha Khadem, L. J. Carr, E. Isaacs, E. Brett, C. Adams, M. Mishkin, 'Onset of Speech after Left Hemispherectomy in a Nine-Year-Old Boy', *Brain*, 120 (1997), pp. 159–82.

32 Sarah Paterson et al., 'Cognitive modularity and genetic disorders', *Science*, 286 (1999), pp. 2355–7.

10

Taking the Stink Out of Instinct

Patrick Bateson

A Little Dose of Judgement

Evolutionary psychology has a recurring theme. Human activities are organised into modular systems, each of which serves a specific use. (See especially Annette Karmiloff-Smith's chapter on modules.) A strong claim is then made that the modules are adapted to their ancestral uses by the Darwinian evolutionary process of natural selection. John Tooby and Leda Cosmides explain that, 'Complex adaptations are intricate machines that require complex "blueprints" at the genetic level.'[1] Later in their well-known article they suggest that specific 'instincts' for dealing with particular problems proliferated in the course of human evolution.

As I will show, such thinking has something in common with that of the ethologists around the middle of the twentieth century. The ethologists' instincts may be traced back to Darwin and earlier writers. However, most ethologists subsequently gave up their ideas about instinct because they generated so much confusion. Did they give up too easily? Have the evolutionary psychologists rediscovered something which had been mistakenly lost? My answer is 'no'. They rediscovered the malodorous aspects of instinct – the stink of confusion that arose because it had been defined in so many different ways.[2]

Charles Darwin wrote in *The Origin of Species* in 1859:

> I will not attempt any definition of instinct. It would be easy to show that several distinct mental actions are commonly embraced by this term; but everyone understands what is meant when it is said that instinct impels the cuckoo to migrate and to lay her eggs in other birds' nests. An action, which we ourselves require experience to enable us to perform, when performed by an animal, more especially by a very young one, without experience, and

157

when performed by many individuals in the same way, without their knowing for what purpose it is performed, is usually said to be instinctive. But I could show that none of these characters are universal. A little dose of judgment or reason . . . often comes into play, even with animals low in the scale of nature.

The reason why Darwin wisely refused to provide a comprehensive definition was because the concept has so many different dimensions to it. The same is true today. At their simplest, instincts may be nothing more than reflex reactions to external triggers, like the knee-jerk or the baby's sucking of a nipple in its mouth. In more complex forms, they are a series of movements all co-ordinated into a system of behaviour that serves a particular end, such as locomotion or non-verbal communication.

Human facial expressions have characteristics that are widely distributed in people of many different cultures. Although they can be deliberately masked or simulated, the emotions of disgust, fear, anger and pleasure can be read off the face, and indeed the entire body, with relative ease in any part of the world. Towards the end of his life Charles Darwin wrote *The Expression of the Emotions in Man and Animals*, a book that provided the stimulus for observational studies of animal and human behaviour which have continued into modern times. Darwin concluded, 'That the chief expressive actions, exhibited by man and by the lower animals, are now innate or inherited – that is, have not been learnt by the individual – is admitted by every one.' Darwin's descriptions of suffering, anxiety, grief, joy, love, sulkiness, anger, disgust, surprise, fear and much else are models of acute observation. He would show to friends and colleagues pictures of people seemingly expressing various emotions and ask them, without further prompting, to describe the emotions.[3] In one case a picture of an old man with raised eyebrows and open mouth was shown to twenty-four people without a word of explanation, and only one did not understand what was intended. In a way that shows both his carefulness and his honesty, Darwin continued, 'A second person answered terror, which is not far wrong; some of the others, however, added to the words surprise or astonishment, the epithets, woful, painful, or disgusted.' His extensive correspondence with travellers and missionaries convinced him that humans from all round the globe expressed the same emotion in the same way. Subsequently, he gathered an enormous archive of photographic records of human expressions in different cultures with many differing forms of economy. The similarities in, for example, the appearance of the smile or the raised eyebrows are striking. The cross-cultural agreement in the interpretation of complex facial expressions is also remarkable. People agree about which emotions are being expressed. They also agree about which emotion is the more intense, such as which of two angry people seems the more angry.[4] Nuances of expression are another matter and the cultural differences in non-verbal communication involving the face raise questions to which I shall return.

At their most complex, instincts are thought to provide the basis by which the individual gathers particular types of information from the environment in the course of learning. The acquisition of language by humans is such a case. Children acquire words and the local rules of grammar from the adults around them, but the way they do so is often thought to be shared by all humans. Therefore, the underlying process is believed to be inherited, internally motivated and adaptive.[5] It is obvious that the differences in spoken language between a French person and a German are not due to genetic differences. Therefore, apart from the act of speech itself, the proposed universal and instinctive characteristics of all humans are not going to be discovered in the surface organisation of such behaviour.

By the mid-twentieth century, those studying human behaviour, particularly in the United States and the Soviet Union, found the late nineteenth-century hereditarian notions offensive. In America the ideology of individualism suggested that everybody could be instrumental in their own route to personal success, while in the Soviet Union the line was that everybody could learn to co-operate and their individualism be subordinated to the common good – and indeed the interests of the state. Behaviourist psychology in the United States and Pavlovian psychology in the Soviet Union provided just the support that was needed for the respective local ideologies. Both schools of thought placed heavy emphasis on the role of learning in the development of behaviour. A new reductionism was born, namely environmental determinism.

The Rise of Ethology

In Western Europe instinct had once again became a focus of intense interest during the middle part of the twentieth century among biologists who were studying the behaviour of animals. Their interest was rekindled by the writings of Konrad Lorenz.[6] He was born in 1903, the son of a successful Austrian surgeon, who tolerated the menagerie of animals kept by Konrad from when he was a small boy. After he qualified in medicine, at his father's insistence, he worked for a PhD in comparative anatomy, meanwhile maintaining his interest in the behaviour of the animals that he kept at the family home at Altenberg. His father remained tolerant of his enthusiasms and supported him throughout the 1920s and 1930s when he had no paid post. During this period his reputation as a scientist was growing and, by the 1930s, he was forging a new theory of instinctive behaviour with his Dutch friend, Niko Tinbergen. In 1973 he, Tinbergen and Karl von Frisch shared the Nobel Prize. Together, Konrad Lorenz and Niko Tinbergen came to be regarded as the founders of ethology – the biological study of animal behaviour.

Niko Tinbergen came from an intellectual family which spawned two Nobel Prize winners: his brother won the prize for economics. Niko Tinbergen moved from Holland to Oxford after the Second World War. In 1951 he published an influential book, *The Study of Instinct*, which set out the main findings and ideas of ethology. Tinbergen's method was to study animals in their natural environments and he was particularly skilled at conducting simple but elegant experiments in the field. Lorenz, on the other hand, preferred to keep animals in his home, where he could more readily observe them.

Lorenz had been struck by how behaviour patterns that had looked so appropriate in the natural worlds to which the animals had been adapted looked so odd when seen out of their normal context. A few days after hatching, a hand-reared duckling touches with its bill a pimple above its tail and then wipes its bill over its down. Yet this pimple, which becomes an oil-producing gland in the adult, is not yet functional and the duckling would normally be oiled by the feathers of its brooding mother. Observations such as this led Lorenz to conclude that behaviour patterns which were well adapted by evolution to the biological needs of the animal are qualitatively distinct from behaviour acquired through learning. Such evidence led Lorenz to contest fiercely what he regarded as the orthodoxies of American behaviourist psychology, with its almost exclusive emphasis on learning.[7]

Lorenz, with his academic training in comparative anatomy, believed that behavioural activities could be regarded like any physical structure or organ of the body. They had a regularity and consistency that related to the biological needs of the animal, and they differed markedly from one species to the next. But while Lorenz was a forceful advocate of the concept of instinct, he certainly did not deny the importance of learning. On the contrary, he gave great prominence to developmental processes by which animals formed their social and sexual preferences. He saw such learning processes as being under the control of what he referred to as the 'innate school marm'. This metaphor represented the highly regulated acquisition of information from the environment just when it is most adaptive for the animal to get it. Lorenz thought of instincts, whether they organised behaviour directly or were the mechanisms that changed behaviour through learning, as inherited neuronal structures which remained unmodified by the environment during development. Behaviour resulting from learning was seen as being separately organised in the brain from the instinctive elements.

Later I shall take a critical look at the initial conception of Lorenz and Tinbergen about instinct. However, it must be remembered that their views were based on many compelling observations of animals' behaviour in the natural environment. These have been added to by a wealth of evidence in subsequent years and many examples of courtship, defensive behaviour, specialised feeding methods, communication and much else are familiar both to researchers and to a wider audience by way of brilliantly made television films. Some behaviour patterns are highly stereotyped in their form. Others are expressed flexibly across a wide range of environmental conditions. The web-making abilities of spiders are like this. The spider building a web explores a potential site, creates a frame of web round the various attachment points, spins radials from the attachment points to the orb, spins more radials to the frame, then spins a spiral from the orb to the outside, and finally another spiral from the frame to the orb. It is an exquisite structure adjusted to the site in which it is built.[8]

Complex and co-ordinated behaviour patterns may also develop without practice. Birds, for example, can usually fly without prior experience of flying. In one experiment, young pigeons were reared in narrow boxes that physically prevented them from moving their wings after hatching. They were then released at the age at which pigeons normally start to fly. Despite having had no prior opportunities to move their wings, the pigeons were immediately able to fly when released, doing so almost as well as the pigeons which had not been constrained.[9] In a similar way, European garden warblers which have been hand-reared in cages nevertheless become restless and attempt to fly south in the autumn – the time when they would normally migrate southwards. The warblers continue to be restless in their cages for about a couple of months, the time taken to fly from Europe to their wintering grounds in Africa. The following spring they attempt to fly north again. This migratory response occurs despite the fact that the birds have been reared in social isolation, with no opportunities to learn when to fly, where to fly or for how long.[10]

Certain aspects of human behavioural development recur in almost everybody's life despite the shifting sands of cultural change and the unique contingencies of any one person's life. Despite the host of genetic and environmental influences that contribute to behavioural differences between individuals, all members of the same species are remarkably similar to each other in many aspects of their behaviour – at least, when compared with members of other species. All humans have the capacity to acquire language, and, discounting extreme circumstances, the vast majority do. With few exceptions, humans pass the same developmental milestones as they grow up. Most children have started to walk by about eighteen months after birth, have started to talk by around two years and go on to reach sexual maturity before their late teens. Individual differences among humans seem small when any human is compared with any chimpanzee. The evidence suggesting that the development of behaviour had many of the same characteristics as the development of the body led to popular arguments about the supposed human instincts for holding territory, cheating and aggression.

The Critique of Instinct

Despite all the empirical evidence that some elements of behaviour can develop without opportunities for learning, the ethologists' notion of instinct attracted strong criticism in the 1950s from a group of American comparative psychologists who studied animal behaviour. A key figure was Ted Schneirla, who worked at the American Museum of Natural History in New York and who was famous for his work on the behaviour of ants. The attacks on instinct were in part motivated by Lorenz's pre-war acceptance of the ideology of the Third Reich. Lorenz had swum with the tide as a member of the Nazi party and in 1940 wrote an article whose infamous ideas were to dog him for the rest of his life. He detested the effects of domestication on animal species and thought (without any evidence) that humans were becoming victims of their own self-domestication. Having got 'our best individuals to define the type-model of our people', the unfortunates who deviated markedly from such a model should, he suggested, be prevented from breeding. Lorenz's wish to rid humanity of what he regarded as impurity matched only too well Hitler's Aryan supremacist dream.

The mix of ethics, science and politics which lay behind the assault on Lorenz's ideas about instinct hit home, and the critics laid out a quite different agenda for studying behavioural development. A leading protagonist in the debate was Danny Lehrman, a brilliant, ebullient and enormously articulate man who later founded the Institute of Animal Behavior at Rutgers University in New Jersey.[11] Lehrman's critique of Lorenz was widely accepted, not least by Niko Tinbergen, not only because he was an anti-fascist but because of the strength of the critique, and Tinbergen's approval shaped the thinking of most English-speaking ethologists. The American psychobiologist Frank Beach referred to this change in thinking as 'the descent of instinct' or, in private, as 'taking the stink out of instinct' – a joke too good not to repeat.[12]

Lehrman insisted that as far as individual development is concerned, the problem has to be expressed differently from the ways in which the earlier ethologists had thought about it. In a later article he put it as follows:

> The problem of development is the problem of the development of new structures and activity patterns from the resolution of the interaction of existing ones, within the organism and its internal environment, and between the organism and its outer environment. At any stage of development, the new features emerge from the interactions within the current stage and between the current stage and the environment. The interaction out of which the organism develops is not one, as is often said, between heredity and environment. It is between organism and environment! And the organism is different at each stage of its development.[13]

The stink (as some of us would see it) of instinct has resurfaced strongly in the late twentieth century, in the writings of sociobiologists and evolutionary psychologists. The 1970s' style sociobiology 'decoupled' (in E. O. Wilson's phrase) individual development from the project to link evolutionary biology and behavioural biology.[14] While few of the 1990s evolutionary psychologists have wished to sink back into a nothing-but-genes position, just where they stand on the developmental project outlined by Lehrman is less clear. How can ideas about modular instinctive behaviour patterns being inherited, internally motivated and adaptive be reconciled with the variable and flexible way in which behaviour develops?

The debate has been confused because the term 'instinct' means remarkably different things to different people.[15] To some, 'instinct' means a distinctly organised system of behaviour patterns, such as that involved in searching for and consuming food. As Annette Karmiloff-Smith describes, the different modules of behaviour have been likened, drawing on a metaphor first used by Cosmides and Tooby, to the various tools found on a Swiss army knife.[16] For others, an instinct is simply behaviour that is not learned. Instinct has also been used as a label for behaviour that is present at birth (the strict meaning of 'innate') or, like sexual behaviour, patterns that develop in full form at a particular stage in the life-cycle. Another connotation of instinct is that once such behaviour has developed, it does not change. Instinct has also been portrayed as behaviour that develops before it serves any biological function, like some aspects of sexual behaviour. Instinct is often seen as the product of Darwinian evolution so that, over many generations, the behaviour was adapted for its present use. Instinctive behaviour is supposedly shared by all members of the species (or at least by members of the same sex and age). Confusingly, it has also been used to refer to a behavioural difference between individuals caused by a genetic difference – so instincts are both universal and part of individual differences. The overall effect of the multiple definitions is, to say the least, muddling (see Table below).

Table   The Various Meanings of Instinct

---

1 Present at birth (or at a particular stage of development)
2 Not learned
3 Develops before it can be used
4 Unchanged once developed
5 Shared by all members of the species (or the same sex and age)
6 Organised into a distinct behavioural system (such as foraging)
7 Served by a distinct neural module
8 Adapted during evolution
9 Differences between individuals are due to genetic differences

---

Some examples can be found to which most of the defining characteristics of instinct seem to apply. The ways in which mice, rats and guinea pigs clean their own fur are good examples of behaviour patterns that do have most of the defining characteristics of instinct. They seem to justify the view that there is a single coherent notion of instinct. If one defining characteristic of instinct has been found, then the rest will also be found. The duration of the elliptical stroke with the two forepaws which the rodent uses to clean its face is proportional to the size of the species; the bigger the species the longer the stroking movement takes. This is not simply a matter of physics. The bigger-bodied species are not slower in their grooming movements simply because their limbs are heavier; a baby rat grooms at exactly the same rate as an adult rat even though it is a tenth of the size.[17] Moreover, young rodents perform these grooming movements at an age when their mother normally cleans them and before their behaviour patterns are needed for cleaning their own bodies. Rodent grooming is, in other words, a species-typical, stereotyped system of behaviour that develops before it is of any use to the individual. It has most of the defining characteristics which have variously been attributed to instinct. This case is by no means typical, however.

Practical Problems

Many of the theoretical implications of the generalised concept of instinct are difficult to test in practice. Take the definition of instinctive as being unlearned, for instance. To establish experimentally that a particular type of behaviour is not learned requires the complete exclusion of all opportunities for learning. This is harder than it sounds. For a start, it is difficult to draw a clear distinction between experiences that have specific effects on the detailed characteristics of a fully developed behaviour pattern and environmental influences that have more general effects on the organism, such as nutrition or stress. Experiences vary in the specificity of their effects.

Even if all obvious opportunities for learning a particular behaviour pattern are excluded, a major problem remains. This is because animals, like humans, are good at generalising from one type of experience to another. It is therefore difficult to know whether an individual has transferred the effects of one kind of experience to what looks superficially like a quite different aspect of their behaviour. For example, if somebody draws a letter of the alphabet on your hand while your eyes are shut you should still be able to visualise the letter, even though you have not seen it. In doing this you will have demonstrated a phenomenon called cross-modal matching. A striking instance of cross-modal matching has been found in rhesus monkeys. Monkeys were trained to distinguish between tasty and obnoxious biscuits in the dark. The obnoxious biscuits, which contained sand and bitter-tasting quinine, differed in shape from the tasty biscuits. The monkeys quickly learned to select the right-shaped biscuits. When they were subsequently tested in the light, the monkeys immediately reached for the nice biscuits, even though they had never seen them before. They had transferred the knowledge they had acquired from a purely tactile experience – touching the biscuits in the dark – and used it to make a visual choice.[18]

Another pitfall in the quest for instinct is that the developing individual cannot be isolated from itself, and some of its own actions may provide crucial experience that shapes its subsequent behaviour. After they hatch, ducklings exhibit an immediate preference for the maternal calls of their own species. Some elegant experiments by the American psychobiologist Gilbert Gottlieb showed that the ducklings' preferences are affected by them hearing their own vocalisations in the egg before hatching. In other words, their 'instinctive' preference is influenced by the sensory stimulation which they generated themselves. Gottlieb was able to demonstrate this by cutting a window in the egg and operating on the unhatched ducklings, thereby making it impossible for them to produce sounds. These silent birds were less able to distinguish the maternal calls of their own species from those of others. However, if they were played tape recordings of duckling calls, the preference for their own species' maternal call emerged.[19]

A formidable obstacle to proving that a behaviour pattern is not learned is the capacity that animals have to acquire the necessary experience in more than one way. When scientists attempt to isolate an animal from one particular form of experience that is thought necessary for development, the behaviour pattern may nonetheless develop by an alternative route. Cats, for example, can acquire and improve their adult predatory skills via a number of different developmental routes: by practising catching live prey when young, by playing at catching prey when young, by watching their mother catch live prey, by playing with their siblings, or by practising when adult. Hence a kitten deprived of, say, opportunities for play may still develop into a competent adult predator but by a different developmental route.[20]

The demonstrations of cross-modal matching, the impact of self-stimulation on the young animal and the use of different developmental routes to the same end-point all sound notes of warning. It is not as easy as it might seem to demonstrate that a behaviour pattern has not been shaped by some form of experience that has a particular influence on the behaviour.

The Concept Breaks Apart

The various characteristics of instinct do not always hang together so closely as they do in the example of rodent grooming. A central aspect of Lorenz's concept of instinct that unravelled on further inspection was the belief that learning does not influence such behaviour patterns once they have developed. Many cases of apparently unlearned behaviour patterns are subsequently modified by learning after they have been used for the first time. A newly hatched laughing gull chick will immediately peck at its parent's bill to initiate feeding, just as, in the laboratory, it will peck at a model of an adult's bill. At first sight this behaviour pattern seems to be unlearned; the chick has previously been inside the egg and therefore isolated from any relevant experience, so it cannot have learned the pecking response.[21] However, as the chick profits from its experience after hatching, the accuracy of its pecking improves and the kinds of model bill-like objects which elicit the pecking response become increasingly restricted to what the chick has seen. Here, then, is a behavioural response that is present at birth, species-typical, adaptive and unlearned, but nonetheless modified by the individual's subsequent experience.[22]

Essentially the same is true for the 'innate' smiling of a human baby. Human babies who have been born blind, and consequently never able to see a human face, nevertheless start to smile at around five weeks – the same age as sighted babies.[23] Babies do not have to see other people smile in order to smile themselves. Just after birth, sighted human babies gaze preferentially at head-like shapes that have the eyes and mouth in the right places. Invert these images of heads, or jumble up the features, and the newborn babies respond much less strongly to them.[24] Despite these observations sighted people subsequently learn to modify their smiles according to their experience, producing subtly different smiles that are characteristic of their particular culture. Nuance becomes important. The blind child, lacking the visual interaction with its mother, becomes less responsive and less varied in its facial expression.[25] The fact that a blind baby starts to smile in the same way as a sighted baby does not mean that learning has no bearing on the later development of social smiling. Experience can and does modify what started out as apparently unlearned behaviour. Conversely, some learned behaviour patterns are developmentally stable and virtually immune to subsequent modification. The songs of some birds are learned early in life, but these learned songs may be extremely resistant to change once they have been acquired.[26]

The idea that one meaning of instinct, 'unlearned', is synonymous with another, namely 'adapted through evolution', also fails to stand up to scrutiny. The development of a behaviour pattern that has been adapted for a particular biological function during the course of the species' evolutionary history may nonetheless involve learning during the individual's lifespan. For example, the strong social attachment that young birds and mammals form to their mothers is clearly adaptive and has presumably evolved by Darwinian evolution. And yet the attachment process requires the young animal to learn the individual distinguishing features of its mother.

Yet another way in which the different elements of instinct fall apart is the role of learning in the inheritance of behaviour across generations. Consider, for example, the ability of birds such as titmice (the European version of the North American chickadee) to peck open the foil tops of the milk bottles that used to be delivered each morning to the doors of a great many British homes. The birds' behaviour is clearly adaptive, in that exploiting a valuable source of fatty food undoubtedly increases the individual bird's chances of surviving the winter and breeding. However, the bottle-opening behaviour pattern is transmitted from one generation to the next by means of social learning. The basic tearing movements used in penetrating the foil bottle top are also used in normal foraging behaviour and are probably inherited without learning. But the trick of applying these movements to opening milk bottles is acquired by each individual bird through watching other birds do it successfully – that is, by social learning. (How the original birds first discovered the trick is another matter.)[27]

In short, many behaviour patterns have some, but not all, of the defining characteristics of instinct, and the unitary concept starts to break down under closer scrutiny. The various theoretical connotations of instinct – namely that it is unlearned, caused by a genetic difference, adapted over the course of evolution, unchanged throughout the lifespan, shared by all members of the species, and so on – are not merely different ways of describing the same thing. Even if a behaviour pattern is found to have one diagnostic feature of instinct, it is certainly not safe to assume that it will have all the other features as well.

Behaviour cannot be neatly divided up into two distinct types: learned and instinctive. Nevertheless, Lorenz's insight that behaviour has something in common with the organs of the body does have substance. The developmental progression from a single cell to an integrated body of billions of cells, combining to produce coherent behaviour, is astonishingly orderly. Just as animals grow kidneys with a specialised biological function, adapted to the conditions in which they live, so they perform elaborate and adaptive behaviour patterns without any previous opportunities for learning or practice. Particular behaviour patterns are like body organs in serving particular biological functions; their structure was likely to have been adapted to its present use by Darwinian evolution and depends on the ecology of the animal; and they develop in a highly co-ordinated and systematic way. This then brings in the question of natural design – a favourite topic in evolutionary psychology.

Appearance of Design

'Biology is the study of complicated things that give the appearance of having been designed for a purpose,' wrote Richard Dawkins in *The Blind Watchmaker*. Dawkins took the image of the watchmaker from an argument developed by William Paley in the early nineteenth century. 'It is the suitableness of these parts to one another; first, in the succession and order in which they act; and, secondly, with a view to the effect finally produced', wrote Paley about the reaction of someone who contemplates the construction of a well-designed object.

The perception that behaviour is designed springs from the relations between the behaviour, the circumstances in which it is expressed and the resulting consequences. The closeness of the perceived match between the tool and the job for which it is required is relative. In human design, the best that one person can do will be exceeded by somebody who has access to superior technology. If you were on a picnic with a bottle of wine but no corkscrew, one of your companions might use a strong stick to push the cork into the bottle. If you had never seen this done before, you might be impressed by the selection of a rigid tool small enough to get inside the neck of the bottle. The tool would be an adaptation of a kind. Tools that are better adapted to the job of removing corks from wine bottles are available, of course, and an astonishing array of devices have been invented. One ingenious solution involved a pump and a hollow needle with a hole near the pointed end; the needle was pushed through the cork and air was pumped into the bottle, forcing the cork out. Sometimes, however, the bottle exploded and this tool quickly became extinct. As with human tools, what is perceived as good biological design may be superseded by an even better design, or the same solution may be achieved in different ways.

Among those who spin stories about biological design, a favourite figure of fun is an American artist, Gerald Thayer. He argued that the purpose of the plumage of all birds is to make detection by enemies difficult. Some of his undoubtedly beautiful illustrations were convincing examples of the principles of camouflage. However, among other celebrated examples, such as pink flamingos concealed in front of the pink evening sky, was a painting of a peacock with its resplendent tail stretched flat and matching the surrounding leaves and grass. The function of the tail was to make the bird difficult to see![28] Ludicrous attributions of function to biological structures and patterns of behaviour have been likened to Rudyard Kipling's *Just So* stories of how, for example, the leopard got his spots.[29] However, the teasing is only partly justified. Stories about current function are not about how the leopard got his spots, but what the spots do for the leopard now. That is a question testable by observation and experiment.

Not every speculation about the current use of a behaviour pattern is equally acceptable. Both logic and factual knowledge can be used to decide between competing claims. Superficially attractive ideas are quickly discarded when the animal is studied in its natural environment. The peacock raises his enormous tail in the presence of females and he moults the cumbersome feathers as soon as the spring breeding season ends. If Thayer had been correct about the tail feathers being used as camouflage, the peacock should never raise them conspicuously and he should keep them all year round.

Conclusion

Bringing together the ideas of design with those of development means that the adaptiveness of the processes must be examined. This process is salutary. From the standpoint of design, systems of behaviour that serve different biological functions, such as cleaning the body or finding food, would not be expected to develop in the same way. In particular, the role of experience is likely to vary considerably from one behavioural system to another. In predatory species such as cats, cleaning the body is not generally something that requires special skills tailored to local conditions, whereas capturing fast-moving prey requires considerable learning and practice to be successful. The osprey snatching trout from water does not develop that ability overnight. Animals that rely upon highly sophisticated predatory skills, such as birds of prey, suffer high mortality when young as a result of their incompetence and those that survive are often unable to breed for years; this is because they have to acquire and hone their skills before they can capture enough prey to feed offspring in addition to themselves. In such cases, a combination of different developmental processes is required in order to generate the highly tuned skills seen in the adult.

The concept of 'instinct' is very far from being unitary. Rather than being a modular Swiss army knife, it is better likened to a kitchen drawer containing a heterogenous collection of implements with different uses. Evidence that a behaviour pattern, or the developmental process that gives rise to it, serves a current biological function does not constitute evidence that the behaviour pattern in question or the developmental process is unlearned. The notion of design helps to sort out what would otherwise look like a confused jumble of behavioural phenomena and the developmental processes that gave rise to them.

My argument is not, therefore, with the application of Darwinian theory to human behaviour but with its misapplication. Behavioural systems are protean in the sense that they take on many different forms when they have been adapted during evolution to current use as much as when they have not. They may be highly conserved from one generation to the next or highly variable. These aspects of human and animal behaviour reinforce the strong conclusion that evolutionary explanations neither replace nor exclude an understanding of how behaviour develops.[30]

Notes and References

1 J. Tooby and L. Cosmides, 'The Psychological Foundations of Culture' in J. H. Barkow, L. Cosmides and J. Tooby (ed.), *The Adapted Mind* (New York, Oxford University Press, 1992), p. 78.

2 If 'instinkt' is spelt in a jokey Russian way, removing the 'stink' leaves INT. This might stand for Integrated and Naturally Tailored behaviour, Internal Networks of Transaction or . . . Thinking of others is a good game.

3 In fact the pictures were photographs of actors who produced an expression on demand or people whose facial muscles had been stimulated with small electric currents.

4 P. Ekman et al., 'Universals and Cultural Differences in the Judgments of Facial Expressions of Emotion', *Journal of Personality and Social Psychology*, 53 (1987), pp. 712–17.

5 N. Chomsky, *Rules and Representations* (New York, Columbia University Press, 1980).

6 A. Nisbett, *Konrad Lorenz* (London, Dent, 1976).

7 K. Lorenz, *Evolution and Modification of Behavior* (Chicago, University of Chicago Press, 1965).

8 F. Vollrath, M. Downes and S. Krackow, 'Design Variability in Web Geometry of an Orb-Weaving Spider', *Physiology and Behavior*, 62 (1997), pp. 735–43. T. Krink and F. Vollrath, 'Emergent Properties in the Behaviour of a Virtual Spider Robot', *Proceedings of the Royal Society of London B*, 265 (1998), pp. 2051–5.

9 J. Grohmann, 'Modifikation oder Funktionsreifung? Ein Beitrag zur Klärung der wechselseitigen Beziehungen zwischen Instinkthandlung und Erfahrung', *Zeitschrift für Tierpsychologie*, 2 (1939), pp. 132–44.

10 E. Gwinner, 'Circadian and Circannual Programmes in Avian Migration', *Journal of Experimental Biology*, 199 (1996), pp. 39–48.

11 D. S. Lehrman, 'A Critique of Konrad Lorenz's Theory of Instinctive Behavior', *Quarterly Reviews of Biology*, 28 (1953), pp. 337–63.

12 F. A. Beach, 'The Descent of Instinct', *Psychological Reviews*, 62 (1955), pp. 401–10.

13 D. S. Lehrman, 'Semantic and Conceptual Issues in the Nature–Nurture Problem', in L. R. Aronson, E. Tobach, D. S. Lehrman and J. S. Rosenblatt (ed.), *Development and Evolution of Behavior* (San Francisco, Freeman, 1970), pp. 17–52.

14 E. O. Wilson, 'Author's Reply to Multiple Review of "Sociobiology"', *Animal Behaviour*, 24 (1976), pp. 716–18.

15 P. Bateson, 'Are There Principles of Behavioural Development?', in P. Bateson (ed.), *The Development and Integration of Behaviour* (Cambridge, Cambridge University Press, 1991), pp. 19–39.

16 D. S. Wilson, 'Adaptive Genetic Variation and Human Evolutionary Psychology', *Ethology and Sociobiology*, 15 (1994), pp. 219–35.

17 K. C. Berridge, 'The Development of Action Patterns', in J. A. Hogan and J. J. Bolhuis (ed.), *Causal Mechanisms of Behavioural Development* (Cambridge, Cambridge University Press, 1994), pp. 147–80.

18 A. Cowey and L. Weiskrantz, 'Demonstration of Cross-Modal Matching in Rhesus Monkeys, *Macaca mulatta*', *Neuropsychologia*, 13 (1975), pp. 117–20.

19 G. Gottlieb, *Development of Species Identification in Birds* (Chicago, University of Chicago Press, 1971).

20 P. Martin and T. M. Caro, 'On the Functions of Play and Its Role in Behavioral Development', *Advances in the Study of Behavior*, 15 (1985), pp. 59–103.

21 Z. Y. Kuo, *The Dynamics of Behavioral Development* (New York, Random House, 1967).

22 J. P. Hailman, 'The Ontogeny of an Instinct: The Pecking Response in Chicks of the Laughing Gull (*Larus atricilla L.*) and Related Species', *Behaviour Supplement*, 15 (1967), pp. 1–159.

23 D. G. Freedman, 'Smiling in Blind Infants and the Issue of Innate vs Acquired', *Journal of Child Psychology and Psychiatry*, 5 (1964), pp. 171–84.

24 V. Bruce and A. Young, *In the Eye of the Beholder* (Oxford, Oxford University Press, 1998), p. 280.

25 H. Troster and M. Brambring, 'Early Social-Emotional Development in Blind Infants', *Child Care Health and Development*, 18 (1992), pp. 207–27.

26 P. Marler, 'Differences in Behavioural Development in Closely Related Species: Birdsong', in P. Bateson (ed.), *The Development and Integration of Behaviour* (Cambridge, Cambridge University Press, 1991), pp. 41–70.

27 See, for example, D. F. Sherry and B. G. Galef, 'Social-Learning Without Imitation – More About Milk Bottle Opening by Birds', *Animal Behaviour*,

40 (1990), pp. 987–9. In recent years milk has been much less commonly delivered to doorsteps and, when it is, it does not have much fat. So the habit is disappearing.

28 G. H. Thayer, *Concealing-Coloration in the Animal Kingdom* (New York, Macmillan, 1909).

29 S. J. Gould and R. C. Lewontin, 'The Spandrels of San Marco and the Panglossian Paradigm: A Critique of the Adaptationist Programme', *Proceedings of the Royal Society of London*, 250 (1979), pp. 281–8.

30 I am very grateful to Hilary Rose and Steven Rose for their comments on a draft of my chapter and, above all, for organising the marvellously stimulating and friendly meeting which gave rise to this book. I have to confess that this chapter strongly resembles one in *Design for a Life* by myself and Paul Martin. The process of working with him was enormously enjoyable and I am much indebted to him.

11

Beyond Difference:
Feminism and Evolutionary Psychology

Anne Fausto-Sterling

According to the fundamentalist Darwinists, feminism is doomed because it refuses to acknowledge scientific truths about human behaviour. True to a long tradition of feminist bashing, science writer Robert Wright, a popular apologist for evolutionary psychology, compares feminism to communism and other 'ideologies that rested on patently false beliefs about human nature'. Like the communist dinosaur, feminists refuse, he suggests, to acknowledge the scientific truth about human nature.[1]

The dispute between Darwinians and feminists has a long history. Four years after the appearance (in 1871) of Darwin's *The Descent of Man*, Antoinette Brown Blackwell published *The Sexes Throughout Nature*. In it she took Darwin to task. He had, she claimed, misinterpreted evolution 'by giving undue prominence to such as have evolved in the male line'.[2] Only a woman, representing a feminine standpoint, could set the record straight. The Darwin Blackwell chides is one whose theory of sexual selection 'supposes that a male superiority has been evolved in the male line'.[3] While females may by default acquire some of these selected characters, 'the more active, progressive male bears off the palm . . . in development of muscles, in ornamentation, in general brightness and beauty, in strength of feeling and in vigor of intellect. Weighed, measured, or calculated, the masculine force always predominates.'[4] Needless to say, such a viewpoint did not sit well with Blackwell and she offered her own interpretation. The more complex or 'advanced' the organism, she suggested, the greater the division of labour between the sexes. For every special character males evolved, females evolved complementary ones. The net effect, however, leads to sexual equality, to 'organic equilibrium in physiological and psychological equivalence of the sexes'.[5]

Despite her book-length retort, in which she advanced evolutionary arguments why men should clean house and wash the dishes, Blackwell lamented her lack of scientific training. 'I do not underrate the charge of presumption which must attach to any woman who will attempt to controvert the great masters of science . . . But there is no alternative! Only a woman can approach the subject from a feminine standpoint; and there are none but beginners among us.'[6]

In Darwin's two best-known works, *The Origin of Species* and *The Descent of Man*,[7] one encounters list upon list, examples multiplied in myriad ways, all provided to form his account of evolution and natural selection. Natural selection takes advantage of a key feature of the biological world – natural variation. A genetically uniform population cannot evolve because there are no varieties to choose among, none better suited than others to succeed in the game of life. Indeed, humans themselves come in many varieties – both physical and cultural. One great difference between Darwin and Wright's favourite modern evolutionists is that all too often the latter present human males and females as invariant. Beneath a surface of difference, they suggest, are universals that rule all humans regardless of their personal history, genetic make-up and culture of origin. Darwin's writings are, of course, embedded in the language and value systems of Victorian England. In a manner characteristic of most scientists of this period, Darwin intertwined sexual and racial difference (Alfred Russel Wallace, Darwin's contemporary theorist of evolution, was, as a socialist and proto-feminist, the exception – as Ted Benton reminds us). Both, Darwin proposed, originated in large part from sexual selection – the very process which modern evolutionary psychologists call on to account for contemporary sex differences. Interestingly, race, as several contributors to this book note, has disappeared from the evolutionary psychologists' account of natural selection. Thus Wright dons liberal clothing when addressing claims of a genetic basis for social differences between the races[8] while still using EP theory to go for the feminist jugular. But back to Darwin. Men, he wrote, love to compete with one another, 'and this leads to ambition'. Women, though, have greater 'powers of intuition, of rapid perception, and perhaps of imitation . . . faculties [also] characteristic of the lower races, and therefore of a past and lower state of civilization'.[9]

Modern evolutionary psychologists follow Darwin in arguing that females are supposed to have evolved to be more sexually reserved than males. One consequence of such reticence is deep confusion about sexual harassment: what seems like normal sexuality to a male registers in females as traumatic, unwanted attention. Wright, for instance, suggests that our legal systems should make adjustments to these evolution-bound sexual differences. He also proposes that women will never break through the glass ceiling because, biologically, they have less of men's innate ambition and willingness to take the risks necessary for success. This particular version of evolutionary theory implies that affirmative action can only result in hiring or promoting inferior candidates. (Discrimination against equally qualified applicants, it seems, no longer happens.) And, as if that weren't enough, some evolutionary psychologists believe that women did not even evolve their own orgasms; it seems we just got lucky because it was so important for men to seek constant sexual gratification. (That the clitoris and the glans of the penis develop from the same embryonic phallus in the foetus is seen as additional support.)[10] Not long after the appearance of E. O. Wilson's *Sociobiology* and *On Human Nature*, anthropologist Donald Symons contrasted the sexually predatory male with the passive female and hypothesised that 'the human female's capacity for orgasm is no more an adaptation than is the ability to learn to read'.[11]

Evolutionary psychologists have a story to tell about human males and females. Males, they suggest, can produce sperm more or less continuously and in large numbers. (In fact these writers often wax eloquent over the prodigiously productive nature of the male reproductive system).[12] In contrast, women produce one egg monthly, and, if impregnated, must incubate it for nine months. Males can start a lot of eggs on the road to babydom (if they can find the females to carry them) and it costs them relatively little to do so in terms of energy production. Without ever offering actual data to show that semen production is less costly than egg production (after all, men don't ejaculate individual sperm free from energy-rich supporting fluid),[13] evolutionary psychology reasons that men must be selected to seek numerous matings with as many partners as possible. Since each ejaculation costs them little, the male strategy is to produce as many children at as low a cost as is possible. In contrast, because they put their eggs in only a few baskets, the females' strategy is to raise fewer, but very high-quality children and see to it that they make it to reproductive age. Thus women should have been selected to be more prudent and choosy about their mates (evolutionary psychologists use the operative and fully loaded word 'coy') because given the large amount of energy and time devoted to each offspring, mating with a low-quality male could have disastrous consequences. A woman, for example, might spend her costly egg and nine months of pregnancy on a child bearing a genetic defect contributed by a sickly male. She must, therefore, have evolved strategies to differentiate between fit and unfit mates. From these basic principles, at least some students of human nature have argued for the evolutionary derivation of many current patterns of human behaviour, such as spousal battering[14] or the Madonna–whore distinction between the 'loose women' men willingly sleep with and the pure ones they want to marry.[15]

Wright summarily dismisses feminists who have demythologised the sorts of claims just described. Giving his critics (including many professional biologists) 'a C− in Evolutionary Biology 101', he opined that 'not a single well-known feminist . . . has learned enough about modern Darwinism to pass judgement on it'.[16] Wright claims to champion the Truth as Science irrevocably teaches it, while painting feminist biologists (and other critics – he does a lot of lumping) as enfeebled thinkers. We are apparently so blinded by political zeal that we have – despite its unassailable truth and our own training as scientists – rejected his view of evolutionary psychology merely because we reject its political message.

What then of the evolutionary origins of these sex and gene differences? David Buss, a prominent academic EP, puts it thus:

> Women face the problem of securing a reliable or replenishable supply of resources to carry them through pregnancy, and lactation . . . especially when food resources were scarce, that is, during droughts or harsh winters. All people are descendants of a long and unbroken line of women who successfully solved this adaptive challenge; for example, by preferring mates who show the ability to accrue resources and to share them.[17]

Buss's story has a certain plausibility. Proto–human females must, indeed, have had the challenge of finding enough nutrition to sustain pregnancy and lactation. But it lacks essential information. Without knowing when the traits of interest became a permanent part of the human lineage, we can know little about the actual environmental variations, little about the degree to which nutritional needs, via an epigenetic system, might have sharpened foraging abilities dormant within some of the genotypes in particular populations, and/or whether systems of natural selection worked to make food utilisation more physiologically efficient. If Buss's selective scenario played, perhaps it fuelled the development of foraging skills, including the ability to hold three-dimensional maps in one's mind's eye – returning even after many years, to a spot which had previously contained a good food source. Certainly Buss can hypothesise that pregnancy and lactation led females to select males who were good providers, just as I can hypothesise that it led females to evolve well-developed spatial and memory skills. We might both be wrong, or right, but without more data and a far more specific hypothesis we have no way of knowing.

There are a lot of data about prehistoric human culture and protohominids and it is appropriate to use them to devise hypotheses about human evolution. It is not unreasonable to ask the hypothesis-builders of evolutionary psychology at least to postulate at what point in human or hominid history they imagine contemporary reproductive behaviours to have first appeared. 'Throughout the Pleistocene' is pretty vague. What is the evidence that it wasn't earlier or later? What, if any, animal systems provide unnamed models? What were the food and predator stresses at that moment? Data on these points can be gleaned from the archaeological and geological record. How did humans respond? Bio-geographic data can be brought to bear on this point. Was there a division of labour during this early period of evolution? Or did gender-based divisions of labour evolve later?[18]

Over how long a period of time did human mating systems evolve? Are they still evolving? For example, the earliest humans living in the heart of Africa certainly did not, as Buss suggests, experience harsh winters. Yet elsewhere within the EP narrative, our savannah past is routinely invoked. How do the events of interest to evolutionary psychology relate in time to the expansion and geographical radiation of human populations? What evidence is there for a long, unbroken line of women? When and where were there genetic bottlenecks during the course of human evolution? How many of them were there? The use of molecular evidence to trace human evolution has created a great deal of ferment during recent years.[19] It would be welcome if evolutionary psychologists were to pay attention to the specific claims of this new research in their theory building. Which evolutionary lines or kinds of adaptive behaviour were lost or selected for? How much of our current gene pool do we have because of genetic drift or geographic isolation, how much because of adaptation and natural selection? Some prior work at least attempts to situate theory making within a time line and a set of postulates about which organisms (chimps? bonobos? *Australopithecus*? *Homo habilis*?) evolved modern human mating patterns.[20] Let's engage in current discussions using the best available knowledge base and the most highly detailed hypotheses available.

Without this greater specification, evaluating competing hypotheses becomes very difficult. For example, given how precarious early human existence must have been, isn't it possible that females realised that no individual male would live long enough or stay healthy enough to provide over a period of years for his offspring? Why isn't it just as likely that the females who passed on more genes to the next generation were the ones who hedged their bets and slept with more than one male? Buss and other evolutionary psychologists engage in what are, in essence, thought experiments, but unless much more carefully specified hypotheses are presented there is no way to know how the postulated starting points relate to the actual starting points.

Table 1 Latour and Strum's Nine Questions

---

1 What are the initial units of evolution? (Genes? Individuals? The family? The species?)
2 Which qualities do the authors think the units possess? (Selfishness? Self-regulation? Harmony? Aggressiveness?)
3 Units with particular qualities enter into relationships with one another. Explicitly, what form do these relationships take? (Exploitative, trade-offs, parasitical, competitive, co-operative?)
4 What time delays are involved in exchanges which take place in the established relationships between fundamental units? (Pre-hominid? Hominid? *Homo*? *Homo sapiens*? Prehistorical? Last week?)
5 What method of measurement can be used to assess answers to questions 1–4? (L and S write, for example, 'it is one thing to state that a baboon behaves as if to improve his reproductive success, but quite another to decide how he can implement this directive when he does not know who his offspring are.' p. 174]
6 In what framework of events is the evolutionary story embedded? [L and S note that most evolutionary stories are logical, but usually not specifically historical.]
7 What agents or causes are said to play a role within the framework of events (e.g. a shift from forest to savannah as a trigger for the evolution of socialness)?
8 What is the stated explicit methodology?
9 What explicit political lessons do the authors of a theory draw?

The development of scientifically sound theories about the evolution of human behavioural patterns and their relationship to contemporary behaviour could emerge from collaborations between social scientists, evolutionists and behavioural biologists. Specifically, those experts in the social studies of science who have been so bitterly attacked in the current science wars have a great deal to offer. One model collaboration, developed in the halcyon days before science studies were taken seriously enough to be attacked, is a paper written by an anthropologist of science, Bruno Latour, and a primatologist, Sharon Strum, who studies baboon behaviour. Latour and Strum[21] devised a set of questions aimed at making specific hypotheses about human evolution. Using their questionnaire they evaluated the quality of the theories constructed by both social scientists and biologists. (All failed the test pretty miserably.) I urge anyone devising theories about evolution and human behaviour to use Latour and Strum's nine questions (see Table 1) to measure the scientific quality of their hypotheses. As they conclude, 'the difficulties of tracing human social origins goes beyond the mere speculative nature of the endeavour. Scientists have not yet come to terms with what makes an account scientific or convincing . . . when scientists are unaware of the mythic character and function of origin accounts . . . the coherence of the scientific account suffers'.

When, instead of hypothesising about past evolutionary events, biologists study evolution in contemporary populations, they collect very particular kinds of data. They monitor food supply, shelter, rainfall, predator levels – sometimes for as long as a decade. At the same time, they follow individual animals as they mate, raise offspring, and die. They observe the animals, use DNA fingerprinting to see who fathered the offspring, and measure changes in animal shape, size, and behaviour over several generations.[22] Evolutionary psychologists also obtain data about contemporary humans and try to reason backwards from what they find. But because their data come from present-day humans, they need to attend especially carefully to human epigenetic systems, that is, to what degree specific behaviours appear in specific environments. Buss writes, 'Women's current mate preferences provide a window for viewing our mating past'.[23] He conducted surveys in both the United States and in thirty-seven cultures world-wide, obtaining a total sample size of more than 10,000. Both men and women rated the importance of eighteen different mate characteristics. In all cases women placed a higher value on men with good financial prospects than vice versa. Buss argues that the present state of affairs resulted from sexual selection. 'Evolution', he writes 'has favoured women who prefer men who possess attributes that confer benefits and who dislike men who possess attributes that impose costs.' He further argues that the evolution of female preference for resource-rich males is ancient and not likely to change. Evidence for the latter claim comes from his observation that the feminist revolution of the 1970s and 1980s did not change this particular preference.

One can look at Buss's data and arguments from several points of view. Social scientists are more than competent to make judgements about data quality, sampling techniques, cross-cultural diversity, propriety of chosen statistical evaluations and such, and indeed some have made pointed criticisms of his research. But evolutionary biologists also have standards for evaluating Buss-like hypotheses. Four, in particular, have been suggested as essential to the acceptance of conjectures about the evolution of human reproductive behaviours.[24] First, of course, is there a good fit between the hypothesis and data? Second, is the evolutionary explanation as good as or better than some proposed alternative? Third, when using questionnaires to obtain data in support of hypotheses about reproduction, do observed or independently documented behaviours correlate with answers on the questionnaire. Fourth, do postulated characters actually relate to reproductive fitness (e.g., does marrying a wealthier man really increase a woman's chance of producing more and fitter children?)? In the case of Buss's hypothesis that, during human evolution, natural selection favoured women who prefer wealthier men, only the first of the four criteria has been reasonably met.

Evolutionary arguments that meet the highest standards of evolutionary science must always hold clear the difference between obtaining data to demonstrate the workings of contemporary selective events, and using contemporary data to devise hypotheses about the past. In his popular writing, Buss often blurs this distinction. For example, he begins his discussion of human female preference for males with financial resources by reference to a field study of a bird called the grey shrike. He cites an elegant experiment in which a field biologist demonstrated that female shrikes preferred males with larger caches of food. The study shows sexual selection at work in a contemporary bird population. Buss then moves from his account of shrike behaviour to imagine a scenario that might have taken place during early human evolution. For female preference for richer mates to have evolved, he stipulates that prehistoric men would have had to be able to accrue and control resources, that different men would have had different resource levels and that there would have been an advantage to monogamy for the female. (As noted above, he never specifies just when during human evolution this might have been going on, so we cannot use the archaeological record to evaluate his assumptions.) These conditions, he feels, are easily met among humans, and he reaches back to the contemporary world to grab as an example a Donald Trump or some Rockefeller or other. He then returns to 'women over evolutionary history' and then back again to contemporary studies of female preference.

What we have, in the end, is a mishmash of argument in which often very beautifully done contemporary studies of mating behaviours in animals are thrown in with far less elegant surveys of contemporary human behaviour. The latter are then combined with unsubstantiated but plausible postulates about some unspecified earlier period of human evolution in which contemporary behaviour might have had its origin. I do not argue that it is wrong to think about the evolution of human behaviour. Rather, one must do it using the high standards of the best studies of behavioural evolution in animals. And if one is going to build hypotheses about prehistoric evolution, then, too, one must use the standards of the field and the rich, albeit imperfect, information already obtained from the fossil record (a point echoed in the chapters by Ted Benton, Stephen Jay Gould and Tim Ingold).

Let us accept that males and females are likely to have evolved different approaches to courting, mating and infant care. Furthermore, those approaches may come into direct conflict. Nonetheless, until recently the vast majority of research on sexual selection has recorded and theorised male behaviours while ignoring, undervaluing and presenting cardboard caricatures of female activities. For example, male dominance was once thought to structure the societies of primates such as African baboons. Females were believed to follow meekly after males, who struggled with one another for leadership and mates. Similarly, the evolution of striking weapons such as antlers or huge canines was attributed to male competition, with the female passively accepting the winner of the tournament as her mate. Based on such examples, a model of human behaviour driven by male competition, male power-plays and often male violence seemed like a natural (in many senses of the word) conclusion.

Over the last two decades, however, ideas about animal sexual behaviour and the evolution of sexual differences have undergone a revolution. During the 1970s women flooded into the field of animal behaviour – especially the study of primates.[25] The new feminism gave them a new way of viewing the world. Most dramatically they began to carefully watch the behaviour of female animals in the field – with astonishing results. For example, they found that female kin groups are responsible for determining much of the social lives of baboons.[26] Why were earlier observers[27] 'unable' to see what today seems obvious? It is possible that their a priori notions about sex roles hindered their abilities to observe. It was not the feminists who were blind to the scientific truth. Rather, their male-biased predecessors made one-sided observations that led them to lopsided accounts of sexual difference.

Once the point had been gained that females actively created their own social environment and that they limited and controlled male behaviour as much as vice versa, the curtains opened wide. Field biologists responded to the criticism that they spent too much time observing male behaviour while making too many a priori assumptions about the nature of female behaviour. The results were stunning. Take paternity, for example. For many years the dogma – and the language – held that males 'sired' offspring. Females, be they humans, blackbirds or butterflies, scientists believed, just lay back and thought of England. A female red-winged blackbird, for instance, selected a territory held and defended by an appropriately macho male. According to the theory, he protected her and the nest from predators while she dutifully laid the eggs, gathered scrumptious insects for his offspring and kept house. End of story – or so anyone taking an animal behaviour class from the 1950s through the 1980s would have learned.

Even for blackbirds, however, life is not so simple. Using DNA technology, scientists made an intriguing discovery: while the hatchlings in a blackbird nest all had the same mother, several different males had got into the paternity act. Having a female on 'his' territory, even copulating with her, turned out to offer no guarantee that a male would achieve paternity. The females seem to be running this show. The blackbird results turn out to hold up in a lot of other species as well. Although the actual rates of offspring fathered by interloping males differ, the message is clear: female behaviour can determine the path of evolution and their activities are every bit as varied, dynamic and complicated as that of males.[28]

The focus, as with Darwin himself, is on variation. A key feature of human evolution was the expansion of the trait of developmental flexibility, leading to the ability to adapt behaviour to context. Careful observations of wild or semi-wild primate populations have yielded some fascinating discoveries about male and female sexuality. Patient observations over many years and several generations, noting which individuals mated, and with whom, which individuals initiated sexual encounters and the consequences of each choice refuted the dogmatic assertions that females don't really want sex, females don't gain from exuberant sexual behaviour, and that females do best by choosing their mates prudently and with discrimination.

In *Female Choices: The Sexual Behavior of Female Primates* Meredith Small[29] found evidence of sexually active female primates whose behaviour is far from coy. In many – if not most – species of monkeys and apes, females are having a lot more sex than seems necessary to produce their usual one or two offspring per year. Some females solicit sex by approaching a male and pushing their hind ends in his face – hardly 'coy' behaviour. Others initiate a game of sexual tag, running to a male, slapping him and retreating for a short distance. The traditional view holds that males should always be ready, willing, and able, while females – good Victorians all – need persuasion, either gentle or otherwise. How can such a viewpoint account for these sexually pushy female apes and monkeys? There is no widely accepted answer, but Small suggests that the immediate pleasure all primates derive from sex can explain the pattern. Others propose that by copulating with several males, a female can disguise the paternity of her offspring, perhaps gaining protection, or at least freedom from harassment, from multiple males each of whom hopes he is the father. Whatever the explanation, Small's work lays to rest the myth that females don't want, don't need and don't get any recreational sex.

Whether for finned fish or legged people, evolutionary models of behaviour are almost never as static as Wright suggests. Indeed, as Tim Ingold's chapter indicates, legged people do not even walk alike. If one imagines that, as do contemporary humans, protohumans almost certainly found themselves in a variety of different environments, the idea of a species-typical set of reproductive behaviours becomes nonsensical. The logic of natural selection suggests that individuals should vary their reproductive behaviours as a function of the environments in which they find themselves. This way of looking at the evolutionary process places the individual in the environment at the centre of the picture. Rather than blaming the victim or her hapless genes, such an evolutionary perspective finds fault with her surroundings. And these, of course, can change.

The essential point is that, in animals and humans alike, male–female interactions around sex and the rearing of offspring are variable matters. Depending on their environments, both sexes can exhibit a wide range of behaviours. Changing the environment can change a set of behaviours. These conclusions contrast sharply with the ideas championed by Wright. He first concedes that calling a behaviour natural does not brand it as forever unchangeable. But what his right hand gives, his left hand takes away. 'People', he writes, '. . . aren't malleable enough to create a society of perfect behavioural symmetry between men and women. Some changes simply can't be made, and others will come only at some cost.'[30] Wright does not reveal the costs, but Wilson's *On Human Nature* does. He considers the price of gender equity and concludes that sexual equality can only be reached by employing extreme social repression. In other words, he sets a price on malleability's head and tells us that this is too dear for most of us in a democratic society to be willing to pay.

This hard-wired view of the inflexibility of human social arrangements flies in the face of the evidence of the plasticity of behaviour.

Even if Buss, Wright and others are 100 per cent on target about the selective forces that led to our current sex/gender systems, broad sweep evolutionary arguments tell us little about specific mechanisms. In the evolutionary psychologist's scenario, individual females who learned to recognise high-resource males survived and reproduced more frequently than those who did not. But what, precisely, were the recognition mechanisms that evolved? Again, one can imagine a variety of possibilities. There might be something about the physique or physiognomy of high-resource males that females could spot. Or perhaps something a lot more indirect and potentially transformative of the human or hominid way of life happened. Perhaps women who talked a lot with other women could gather information through social and cultural networks. Perhaps, in this scenario, what evolved was the ability to gossip and trade information about nearby males (Barkow[31] also discusses the evolution of gossip).

The result might be the evolution of elaborate cultural mechanisms, not some built-in hard-wired unchangeable brain response. It is precisely the plasticity of the ways in which genetic mechanisms can respond to environmental differences – the so-called norm of reaction – that this hard-wired approach ignores. It presumes that, in the absence of empirical measures, one can predict for any or all environments the phenotype produced by a particular gene. Yet both physical phenotypes and behavioural phenotypes are plastic. Take as an example eastern bluebirds. While almost never physically aggressive to bluebird females, the males will defend scarce nesting sites, which they broker in exchange for sex. But what happens when there are plenty of places to nest, for example, when people start lining fields with nesting boxes? The female doesn't need to trade sex for lodging. Under these circumstances male bluebirds change strategies. They help females feed their babies. One approach may be particularly effective in locations or seasons when food is scarce; while at other times and places, females may be able to fend for themselves. In other words, both male and female bluebirds exhibit behavioural plasticity. Under plentiful conditions, the ornithologist Patricia Gowaty, studying bluebirds, predicts that females will mate with whichever and how many males strike their fancy.[32] In these cases, it makes more sense to think of the plasticity, rather than the specific behaviour, as being under genetic control. In fact, geneticists studying animals and plants have amassed experimental data showing that plasticity is a trait under genetic control, and can evolve via natural selection. Among ecologists and quantitative geneticists, the evolution and genetics of plasticity is a very hot topic.[33] Yet neither the concepts of the norm of reaction or phenotypic plasticity have appeared as a serious part of evolutionary theories about human sex differences.

The discussions of sex differences that one reads in various settings, in both scholarly texts and in the media, frequently slip from one category of biological explanation to another. This slippage makes it difficult to assess the strengths and limitations of particular knowledge claims. One finds oneself simultaneously coping with reductionist and evolutionary explanations for behaviour, which address rather different biological questions and demand different types of proof. To have intelligent discussions and arguments about the role of biological difference in the genesis of gender difference, we must attend to what level of explanation is being offered. Those with whom we debate these questions need to be held to a higher standard of explanatory clarity than has hitherto been offered.

Evolutionary explanations of difference often entail elegant theories based on very partial knowledge of contemporary cultures and on analogies from animals, but without any foundation in the specific history of human evolution. There are no studies of human evolution comparable to those on red deer[22] or chimpanzees.[34] And the two species of chimps, for example, have strikingly different mating systems. Which shall we choose as our model female? Females of the better-known chimp species have an associated pattern of hormones and copulation, but the bonobo female has sex constantly with both males and females and apparently uses sex not just for reproduction, but as a medium of social mediation.[35] Evolutionary explanations of human sex differences usually ignore an entire literature on norms of reaction and phenotypic plasticity. Using this strong and interesting literature in basic genetics and ecology could lead to a very different kind of story-telling.

The brand of evolutionary psychology championed by EP presents a cardboard version of both animal and human females. It is only slightly kinder to men. They may find appealing the image of a Lothario, always ready for sexual conquest, willing – albeit with regret – to use justifiable violence to hold on to his sexual investments and fired with ambition in the workaday world. Some men, however, may consider this cut-out to be a bad fit, while others may well resent Wright's flip reference to 'the average beer-drinking, two-timing, wife-beating lout'.[36] Many men may wish for an account of human existence that permits them a little flexibility, and that offers them some options for political change. But there is a better vision, based on a knowledge of biology both broad and deep, which may permit development of an alternative to the stultifying politics of the status quo.

Notes and References

1 Robert Wright, 'Feminists, Meet Mr Darwin', *New Republic* (28 November 1994), pp. 34–46. Quote on p. 34). As a senior editor of *New Republic* Wright had extraordinary media access. His was the cover story of the 28 November issue; he also had a piece in the *New Yorker* and a radio interview on PBS.

2 Antoinette Brown Blackwell, *The Sexes Throughout Nature* (1875; reprinted Westport, Conn., Hyperion Press, 1976), p. 20.

3 Ibid., p. 18.

4 Ibid.

5 Ibid., p. 58.

6 Ibid., p. 22.

7 Charles Darwin, *The Origin of Species by Means of Natural Selection* (London, John Murray, 1876); Charles Darwin, *The Descent of Man* (London, Murray, 1871).

8 See Wright's editorial in *New Republic* (1 January 1995), p. 6.

9 *The Descent of Man*, pp. 326–7.

10 Donald Symons, *The Evolution of Human Sexuality* (Oxford, New York, Oxford University Press, 1979), writes, 'I have already suggested that the potential for female orgasm can be understood as a byproduct of selection for male orgasm' (p. 94). In contrast, Buss notes that orgasm increases sperm retention; evolution, then, ought to select for men who stimulated female orgasm during or just following intercourse. David M. Buss, *The Evolution of Desire: Strategies of Human Mating* (New York, Basic Books, 1994), pp. 75–6. In Buss's version of evolutionary psychology women have much more agency than they do in Symon's. As we will go on to show, not all evolutionary psychologists think alike.

11 Symons, *The Evolution of Human Sexuality*, p. 312.

12 See also Martin's account of the metaphors of sperm and egg in Emily Martin, *The Woman in the Body: A Cultural Analysis of Reproduction* (Boston, Beacon, 1987).

13 Even at the time the idea was first put forth, some scientists pointed out the unlikelihood of the basic assumption that sperm had no cost. See D. A. Dewsbury, 'Ejaculate Cost and Male Choice', *American Naturalist*, 119 (1982), pp. 601–10, and D. Austin and D. A. Dewsbury, 'Reproductive Capacity of Male Laboratory Rats', *Physiology and Behavior*, 37 (1986), pp. 627–32. For more recent doubts about low-cost sperm see D. R. Levitan and C. Petersen, 'Sperm Limitation in the Sea', *Trends in Ecology and Evolution*, 10 (1995), pp. 228–31; T. L. Karr and S. Pitnick, 'The Ins and Outs of Fertilization', *Nature*, 379 (1996), pp. 405–6; S. Pitnick and T. L. Karr, 'Sperm Caucus', *Trends in Ecology and Evolution*, 11 (1996), pp. 148–51; and references cited by Anne Fausto-Sterling, 'Attacking Feminism Is No Substitute for Good Scholarship', *Politics and the Life Sciences*, 14 (1995), pp. 171–4.

14 Robert Wright, *The Moral Animal* (New York, Pantheon, 1994), pp. 354–5.

15 Ibid., p. 31.

16 Wright, *New Republic*, pp. 36–7. This passage specifically attacks *Myths of Gender: Biological Theories About Women and Men* by Anne Fausto-Sterling (New York, Basic Books, 1992) and Carol Tavris, *The Mismeasure of Woman* (New York, Simon & Schuster, 1992).

17 David M. Buss, 'Psychological Sex Differences: Origins Through Sexual Selection', *American Psychologist*, 50 (1995), pp. 164–8.

18 Lila Leibowitz, *Females, Males, Families: A Biosocial Approach* (North Scituate, Duxbury Press, 1978).

19 Francisco Ayala, 'The Myth of Eve: Molecular biology and human origins', *Science*, 270 (1995), pp. 1930–6. Elizabeth Culotta, 'Asian Hominids Grow Older', *Science*, 270 (1995), pp. 1116–17; Robert L. Dorit, Hiroshi Akashi and Walter Gilbert, 'Absence of Polymorphism at the ZFY Locus on the Human Y Chromosome', *Science*, 268 (1995), pp. 1183–5; Ann Gibbons, 'Rewriting – and Redating – Prehistory', *Science*, 263 (1994), pp. 1087–8; Ann Gibbons, 'The Mystery of Humanity's Missing Mutations', *Science*, 267 (1995), pp. 35–6; Ann Gibbons, 'The Peopling of the Americas', *Science*, 274 (1996), pp. 31–3; Michael F. Hammer, 'A Recent Common Ancestry for Human Y Chromosomes', *Nature*, 378 (1995), pp. 376–8; Alberto Piazza, 'Who Are the Europeans?', *Science*, 260 (1993), pp. 1767–9; S. A. Tishkoff et al., 'Global Patterns of Linkage Disequilibrium at the CD4 Locus and Modern Human Origins', *Science*, 271 (1996), pp. 1380–5.

20 Lila Leibowitz, *Females, Males, Families: A Biosocial Approach* (North Scituate, Duxbury Press, 1978); Nancy Makepeace Tanner, *On Becoming Human* (Cambridge, Cambridge University Press, 1981); Linda Fedigan, 'The Changing Role of Women in Models of Human Evolution', *Annual Review of Anthropology*, 15 (1986), pp. 25–66.

21 B. Latour and S. C. Strum, 'Human Social Origins: Oh Please, Tell Us Another Story', *Journal of Social and Biological Structures*, 9 (1986), pp. 169–87.

22 T. H. Clutton-Brock, F. E. Guinness and S. E. Albon, *Red Deer: The Behavior and Ecology of Two Sexes* (Chicago, University of Chicago Press, 1982); Nicholas B. Davies, *Dunnock Behavior and Social Evolution* (New York, Oxford University Press, 1992).

23 David M. Buss, *The Evolution of Desire* (New York, Basic Books, 1994), quotes from pp. 23, 21.

24 Kim Wallen, 'Mate Selection, Economics and Selection', *Behavioral and Brain Sciences*, 12 (1989), pp. 37–8.

25 Donna Haraway profiles a number of important female primatologists in *Primate Visions: Gender, Race and Nature in the World of Modern Science* (New York, Routledge, 1989). See also Meredith Small (ed.), *Female Primates: Studies by Women Primatologists* (New York, A. R. Liss, 1984).

26 Shirley Strum, *Almost Human: A Journey into the World of Baboons* (New York, Random House, 1987).

27 Irven de Vore, *Primate Behavior: Field Studies of Monkeys and Apes* (New York, Holt, Rinehart & Winston, 1965).

28 For a review of this literature see Patricia Adair Gowaty, 'Field Studies of Parental Care in Birds: New Data Focus Questions on Variation Among Females', in C. T. Snowdon and J. S. Rosenblatt (ed.), *Advances in the Study of Behavior*, 24 (1995).

29 Meredith Small (ed.), *Female Choices: The Sexual Behaviour of Female Primates* (New York, Cornell University Press, 1984).

30 Wright, *New Republic*, pp. 44–5.

31 Jerome H. Barkow, 'Beneath New Culture Is Old Psychology: Gossip and Social Stratification', in Jerome H. Barkow, Leda Cosmides and John Tooby (ed.), *The Adapted Mind: Evolutionary Psychology and the Generation of Culture* (New York, Oxford University Press, 1992), pp. 627–37.

32 Patricia Adair Gowaty, 'Field Studies of Parental Care in Birds: New Data Focus on Variation Among Females', in C. T. Snowdon and J. S. Rosenblatt (ed.), *Advances in the Study of Behavior* (New York, Academic Press, 1996), pp. 476–531; Patricia Adair Gowaty and William C. Bridges, 'Nestbox Availability Affects Extra-Pair Fertilizations and Conspecific Nest Parasitism in Eastern Bluebirds, *Sialia sialis*', *Animal Behavior*, 41 (1991), pp. 661–75; Patricia Adair Gowaty and William C. Bridges, 'Behavioral, Demographic and Environmental Correlates of Extrapair Fertilizations in Eastern Bluebirds, *Sialia sialis*', *Behavioral Ecology*, 2 (1991), pp. 339–50.

33 Robert A. Newman, 'Adaptive Plasticity in Development of *Scaphiopus couchii* Tadpoles in Desert Ponds', *Evolution*, 42 (1988), pp. 774–8; Richard Gomulkiewicz and Mark Kirkpatrick, 'Quantitative Genetics and the Evolution of Reaction Norms', *Evolution*, 46 (1992), pp. 390–411; Samuel M. Scheiner, 'Genetics and the Evolution of Phenotypic Plasticity', *Annual Review of Ecology and Systematics*, 24 (1993), pp. 35–68; Carl D. Schlichting and Massimo Pigliucci, 'Gene Regulation, Quantitative Genetics and the Evolution of Reaction Norms', *Trends in Ecology and Evolution*, 9 (1994), pp. 154–68; Mary Jane West-Eberhard, 'Phenotypic Plasticity and the Origins of Diversity', *Annual Review of Ecology and Systematics*, 20 (1989), pp. 249–78.

34 Jane Goodall, *The Chimpanzees of Gombe: Patterns of Behavior* (Cambridge, MA, Harvard University Press, 1986).

35 Amy Randall Parish, 'Sex and Food Control in the "Uncommon Chimpanzee": How Bonobo Females Overcome a Phylogenetic Legacy of Male Dominance', *Ethology and Sociobiology*, 15 (1994), pp. 157–79; Frans B. M. de Waal, 'Bonobo Sex and Society', *Scientific American* (March 1995), pp. 82–8.

36 Wright, *New Republic*, p. 44.

12

Different Strokes: Beyond Biological Determinism and Social Constructionism

Tom Shakespeare and Mark Erickson

Evolutionary psychology presents an important challenge to both the biological and the social sciences. Part of meeting that challenge is to provide evidence to demonstrate the fallacies, contradictions and limitations of those writing under the banner of evolutionary psychology. Yet there is also a duty to respond more positively, in order to provide alternative, sophisticated, but equally accessible models for understanding human behaviours and human societies. Particularly, it is important to move beyond the tradition of 'two cultures' and to seek ways of connecting social with biological insights and research. Evolutionary psychology has attempted to provide such a synthesis, but only by collapsing much of the social world into an ultra-Darwinian model in which biological imperatives predominate. But an adequate account would not privilege either the biological or the social end of the polarity. Instead it would seek to transcend the dichotomy, moving beyond the inadequacies of dualism. This chapter offers ways forward for understanding the complex inter-relations which operate in the everyday world of disabled people. We draw on the lived experience of difference to develop alternative approaches which avoid the extremes of biological determinism and social constructionism that have bedevilled so many attempts to theorise difference. As against the myth of the Standard Social Science Model – the SSSM – caricatured by evolutionary psychology (see Hilary Rose's and Ted Benton's chapters) we explore how better understandings of impairment and disability can contribute to the construction of new biosocial models.

Either/Or

This dualistic reasoning constantly seeks to push explanation into either the biological or the social. Are disabled people victims of their impaired bodies or victims of a society that discriminates against them? Is ME (myalgic encephalomyelitis) an organic illness or a psychiatric condition? Is RSI (repetitive strain injury) the result of an 'eggshell personality', as the courts ruled in the case of the *Financial Times* journalists, or of excessive keyboard use? Are men and women gay or lesbian because they were born with a particular gene or because they choose to have relations with the same sex? Judgements such as these are both discomforting for the individuals directly affected, and of considerable contemporary significance. Resolving the definitional questions embedded in these dichotomies is critical not only for individuals trying to make sense of their lives, but also for legal and health-care professionals responding to the challenges for justice and care. Yet a good part of the difficulty is that these debates – between nature and nurture, between choice and determinism and between external definition and self-construction – are rooted in binary oppositions, which are themselves invalid or unhelpful in understanding the complexities of human experience.

Compounding these difficulties is the character of current theories. There appears to be a tendency to move towards generalised perspectives, which extrapolate all experiences towards polar opposites and have great difficulty in explaining personal or localised circumstances. One set of such answers is provided by the advocates of evolutionary psychology, behavioural genetics, or other accounts of social experience which prioritise biological explanation. In these approaches it is possession of a specific set of genes, or a particular configuration of the hard-wired brain, constituted via evolutionary mechanisms, which explains any given social phenomenon. This includes not just individual lives, but the broader status of concepts as varied as love and desire, altruism, culture and mental health. It is notable that at the end of the twentieth century such explanations have become increasingly popular. As religion and Marxism have lost their salience for many in the West, Darwinism and the new 'biological metaphysics' provide simplistic answers to the question of origins and causes. Moreover, these answers seek to carry with them the aura and legitimacy of natural science in an era when scientific explanation is still the most powerful mode of explanation. Yet, as others in this volume have demonstrated, such approaches are theoretically deeply flawed, offering little that can advance either the life or social sciences.

For much of the twentieth century a competing set of answers has been provided by the social sciences. But the explanatory models used by social scientists, which have emphasised structure, have recently come under question with what is widely spoken of as the literary turn. Whereas modernist sociology, for example, answers questions of origins and causes with reference to structural concepts, such as capital and class – or much more recently patriarchy and gender – or to the solidarity produced by shared religious values, much contemporary poststructuralist social theorising rejects such explanations. Increasingly, following the lead of French philosophers such as Jacques Derrida and Michel Foucault, this strand of contemporary theory centres on the ways we develop our knowledge of the world, as much as on the world itself.[1] If we can indeed only understand or describe reality by means of language, then language becomes the topic of study. In this poststructuralist/postmodernist perspective, different accounts of the world – including those of the natural sciences – become competing stories. Realism, not for the first time, finds itself under profound interrogation from relativism.

Such poststructuralist theorising centres on what people say and think rather than on what they do or experience. For example, in Judith Butler's *Gender Trouble* categories of gender – such as female and woman, masculine and man – become detached from fixed meanings. They cease to be stable notions and can only be understood by reference to the ways in which language constructs the categories of sex.[2] 'Performativity', not the sex gender system or patriarchy, is key.

These new discourses by and large eschew the term social sciences, as that is too redolent of stucture and the categories of modernity; instead they speak of the human sciences. This new development presents us with two problems. First, they are inaccessible. Writers such as Butler – and there are many others – top the polls of incomprehensibility. Theoretical language has become largely removed from everyday language. Second, these accounts are deliberately divorced from the concept of experience, on the grounds that the unproblematic reading off of experience by structuralists is a major source of weakness. However, the price of this focus on language and interpretation is that it is not possible to take into account material conditions. The limitations of poststructuralism are identified by Harriet Bradley, who argues for a critical realism which accepts structure while not reducing it all to class:

> It would be nice if the social world were no more than a contestation, so that, merely by renaming the world, we could change it ... This underestimates the multi-dimensionality of gendered power which has both cultural and material aspects ... Our everyday engagement with the process of defining the world takes place within relationships of power which involve differential control of, and access to, a range of resources, material,

political, cultural and symbolic, including the utilisation of means of force and violence.[3]

Perhaps it is the lack of clarity and relevance of contemporary human sciences which explain why so many turn instead to bestsellers such as *How the Mind Works* and *The Origin of Virtue*. Yet are literary deconstructionist or biological determinist explanations the only possible choices?

It would be more helpful to begin by recognising that phenomena such as disability are intrinsically complicated and multi-dimensional. Bruno Latour tries to move beyond this either/or thinking in his discussion of the dominance of 'hybrids' in the contemporary social world. His hybrids, like Donna Haraway's 'cyborgs', are combinations of nature and culture.[4] He suggests that we must move beyond the nature/nurture argument, to construct a model in which the inevitable dialectic of factors is represented in notions of networks, continuities and skeins, rather than polarities, oppositions and absolutes. We require a juggling habit of mind, to replace the unimaginative tendencies of either/or, with the rich complexities of both/and.

Both/And

To make sense of these complex social phenomena we need to consider both material conditions (including embodiment) and the cultural and social processes in which they are located. This form of accounting draws on sociological insights and makes space for biological process. To demonstrate the reasons why an either/or approach is inadequate, and to highlight the ways in which both/and approaches can be fruitful, we will look at a range of examples.

The traditional definition of disability centres on the biological deficit of disabled people – the limitation of body or mind. People with particular impairments are defined in terms of what is 'wrong' with them, and which 'normal' activities or functions they cannot perform. In this account, disability and disadvantage are a straightforward consequence of having an impairment. This 'medical model' has dominated both academic thinking and social practice in the health and welfare field until recent years. But explaining the exclusion and disadvantage which disabled people experience in terms of their physiological or mental deficits is a form of 'blaming the victim', and serves to remove all responsibility from societal context. It is unhelpful to reduce disablement to an individual, biological attribute, when many disabled people identify collectively as members of a minority group facing oppression in society, rather than as a group of individuals deviating from the bodily 'norm'.[5]

The political movement of disabled people has rejected the 'medical model' and substituted a 'social model' in its place. Just as early feminist writers such as Ann Oakley[6] made a distinction between sex (the biologically given male and female) and gender (the social experience of being a man or woman), so disabled people distinguish impairment (a bodily difference) and disability (the way that society treats people with impairments). For the disability movement the problem is not having an impairment, but being disabled by society. Just as men and women have been regarded differently at different stages of history, so the status and identity of disabled people has changed in different contexts. Of course, this does not have to imply that disabled people form one unitary group: there are as many different ways of being a disabled person as there are of being a woman.[7]

The effect of the changed definition is to relocate the issue to the social and structural level. In doing this it returns to the classical form of explanation of the social sciences but extending the structure to include a new group. Thus the problem of disability might include factors such as: employment discrimination; lack of access to transport facilities or to housing or to public space; poverty; increased exposure to physical and sexual abuse; prejudiced cultural representations; interpersonal attitudes such as paternalism, intrusive curiosity and ridicule, or outright hostility. The growing numbers of researchers in the field of disability studies have found substantial empirical evidence for each and all of these problems.

The problem of disability is as much or more to do with social and cultural processes as it is to do with biology. This is clearly demonstrated by the way in which the experience of people with particular impairments differs depending on the country in which they find themselves: the United States, thanks to the Disabilities Act of 1990, is now a less disabling environment than, say, the United Kingdom. Equally Newcastle, with a largely accessible Metro, is less disabling for travellers with impairments than London, with a largely inaccessible Underground. Moreover, access to wealth and privilege reduces the difficulties of impairment: high-status white males such as President F. D. Roosevelt, Professor Stephen Hawking and actor Christopher Reeve can enjoy a quality of life which paralysed people in different social situations can only dream of.

Disabled activists have flexed their 'sociological imaginations' by making private problems into public issues. They have used a sociological approach to redefine their experience, and identify the barriers which they want removed. Yet recently some disabled feminists have become disenchanted with the social model because of its failure to confront physical impairment and its overriding stress on structural barriers. As Liz Crow suggests:

> Sometimes it feels as if this focus is so absolute that we are in danger of assuming that impairment has no part at all in determining our experiences. Instead of tackling the contradictions and complexities of our experiences head on, we have chosen in our campaigns to present impairment as irrelevant, neutral and, sometimes, positive, but never, ever as the quandary it really is.[8]

While Liz Crow and colleagues like Jenny Morris and Liz French do not deny that society causes many problems, they also feel that their bodies may cause difficulties, and they want any theory of disability to take account of the physical dimension to their lives. They suggest that in developing a social and structural analysis the disability movement has omitted a key facet of their experience.

Here is an example, then, in which taking an absolute position has proved inadequate. First, disability was equated with biology in the medical model, but this did not take account of the variations in the social experience of disability and the steps that could be taken to change the experience of having an impairment. Second, radical disabled people produced the social model. Yet explaining disability purely in terms of social and external factors ran the risk of ignoring the bodily realities of living with illness or impairment. Clearly, a third model is needed, one that takes proper account of both the personal and physical experience of disability, and the social dimensions. It needs, too, to recognise the importance of psychological processes and the cultural patterns and representations which influence the way we think about disabled people and as disabled people. These four dimensions of analysis, inextricably entwined, produce the disability phenomenon which millions of people experience every day.

And/Or

Our example of disability is perhaps straightforward, in that disability is predicated on particular physical bodies operating in discriminatory social environments. Any adequate model of disability has to explain the phenomena with reference to these different dimensions, and we argue that the both/and model is superior to the existing dichotomous either/ or models. But in other situations, the either/or model has to be replaced with an understanding of the ways that categories may arise out of biological difference, but then also become social and cultural possibilities. This results in a range of ways of being which cannot be explained purely by reference to biological processes. Where we saw disability as best explained by a both/and model, here we need to think in terms of phenomena being both biological 'or' social, as well as biological 'and' social.

Robert Hertz was an early twentieth-century anthropologist who wrote an imaginative book called *Death and the Right Hand*. In the first section Hertz shows that death, an inescapably biological process and event, is also a social process and event, by showing how different cultures manage the cultural transition of the person from life to death, and the rituals which surround funeral and burial. However, it is the second section of the book that is most relevant to our argument. Hertz starts off by suggesting that, of the people who are born with a dominant hand, more are right-handed than left-handed. This means that right-handedness tends to predominate culturally: it becomes the norm. However, Hertz argues, many people are not born with one hand dominant. They are born potentially ambidextrous. Yet because they are born into a right-handed culture, they will be taught, and accustom themselves, to use their right hand skilfully.

This explanatory model suggests that while most people are born right-handed, others acquire right-handedness. Some are innately left-handed, other 'left-handers' become right-handed through development. It seems very simple and very obvious. Yet it is a step forward to recognise plasticity in learning and development from the usual tendency of seeing biology and society as deterministically dichotomous. The anthropologist Mary Douglas casts some light on this, arguing that we are a species which tends to categorise and classify.[9] Thus when we see a distinction in nature, then a process of amplification may create a wider pattern of meaning. To go back to handedness, this has in the past had moral overtones: the word 'sinister' derives from the Latin for left.[10]

This form of explanation may be particularly relevent to homosexuality, where there is a fierce debate about particular differences that might exist in the brains or genes of gay men.[11] Many gay people have been resistant to a biological explanation of gayness, suggesting that it is a matter of individual choice and social custom, while others have claimed to have felt different from a very early age. Yet both gays and straights (particularly biologists) have argued that explaining a complex phenomenon like sexuality in terms of brain region or genetic determinism is a biologically mistaken endeavour. Following Hertz, we might suggest that while some people are 'born gay', and some people are 'born straight', a large proportion of people (maybe a majority) are 'born bisexual' (a claim made by Freud, among others), and that this group becomes set into a heterosexual lifestyle because it is the social and cultural norm. Choosing to be gay or bisexual in such a context is to swim against the cultural tide. Homosexuality, then, may be both biological (whether genetic or brain region explanations are invoked) and also the result of social processes. Thus theory needs to take into account that people are gay or bisexual for different reasons, only some of which may be biologically grounded. The sociologist would also want to point out that for all homosexual and bisexual people, social and cultural contexts and values have a major impact on their lives and identities. And, as others in this book have shown, brain regions, even if 'different' from the 'normal' (both terms concealing a moral economy of the body), do not in themselves determine specific developmental paths. (See for instance Annette Karmiloff-Smith's discussion of Williams syndrome). Further, as Steven Rose and Gabriel Dover point out, genes cannot be understood except in relation to the organism as a whole, the processes of development and the particular environment in which development takes place.[12]

Before/After

Yet this discussion still falls into the danger of thinking that the categories 'homosexual' and 'heterosexual' are universal and natural types. However, social and historical research suggests otherwise. Mary McIntosh, Michel Foucault and others have suggested that the category 'homosexuality' is a social invention of the eighteen and nineteenth centuries.[13] Same-sex behaviour is a feature of all cultures, but at a specific historical time homosexuality became a particular identity; as McIntosh puts it, a 'deviant role'. Whereas previously there had been a continuum of sexual behaviour, after this point there was an increasingly firm divide between the heterosexual majority and the homosexual minority.

This tells us something very important about social concepts: in order for people to be labelled, or to identify themselves, the category has to exist in the first place. It has to be discovered or named as a possibility. Before the concept is invented people who display the behaviour associated with that category are not separated out or distinguished. The act may be unobtrusive, or may be lumped together with a variety of other different behaviours, or social changes may make it suddenly visible. After that point, people with the condition or exhibiting the behaviour are seen as a different class of person from everyone else.

The philosopher Ian Hacking calls this invention of social categories 'making up people'. He suggests that 'numerous kinds of human beings and human acts come into being hand in hand with our invention of the categories labelling them'.[14] When identities emerge – perhaps, as in the case of homosexuality, because they become a focus for concern and prohibition – the possibilities for personhood change. This includes both the way that people are allocated to categories and the way that they themselves identify.

Of course, some categories – male and female, right- or left-handed – are more immanent than others. To talk of either discovery or invention is unnecessary, although there undoubtedly came a point in the development of the human species when a realisation of these differences developed. But identities such as homosexuality and conditions such as autism depend on a process of naming and defining in order to become social possibilities. In the case of dyslexia the condition would have remained largely invisible until literacy became a common cultural acquisition and orthography was standardised.

Conditions like autism seem to be universal: across different cultures, the incidence remains fairly stable. Yet before Asberger and Kanner separately identified the specific set of behaviours which indicate the impairment, the category 'autism' did not exist. Debate still rages about the precise diagnostic details and the causes of the condition: some refuse to accept that it has an organic, probably genetic, origin. What is true of autism is even clearer in the case of recently discovered diseases such as Chronic Fatigue Syndrome or ME, and also dyslexia, Attention Deficit Hyperactivity Disorder and other developmental difficulties. In each case there is a debate as to whether the condition actually exists at all; as to whether it is organic, or psychosomatic, or entirely imagined, or an outcome of social labelling. There are also medical and legal arguments about particular cases: will the doctor or employer accept that this adult is genuinely ill? Is this child suffering a specific learning difficulty or are they just lazy or less intelligent?

Chronic Fatigue Syndrome/ME is the archetypal controversial illness. Some advocates of the diagnosis trace the condition back into the past, suggesting that the exhaustion that Charles Darwin, Florence Nightingale and other historical figures continuously suffered was undiagnosed CFS/ME. Recently the literary critic Elaine Showalter has caused a storm of controversy by debunking the concept. She links ME to the *malades imaginaires* from which many people, usually middle-class women, suffered in the nineteenth century. For Showalter, ME is the contemporary equivalent of 'hysteria'. So does CFS/ME exist as an organic condition or is it merely a form of hypochondria or mental illness? Is it biological or social?

Our approach is to suggest an and/or model. CFS may well have a physiological expression, perhaps even indicate an organic dysfunction, but many (perhaps disproportionately middle-class people) have 'jumped on the bandwagon', seeking access to what may be a privileged diagnosis.[15] A recent exchange in the *London Review of Books*, referring to the difference between Elaine Showalter's argument in *Hystories* and the CFS research conducted by Simon Wessely and others, supports this idea. Basing herself on her own experience of a definite organic dysfunction, Sarah Rigby criticised Showalter's claim that ME/CFS is a middle-class myth.The clinician Simon Wessely commented:

> Our work showed that the majority of those who come to a specialist clinic such as the one I run at King's College Hospital do indeed come from the professional classes. Nearly all believe that they are suffering from myalgic encephalomyelitis (ME), and many do not fulfil established international criteria for CFS. An illness which preferentially affects the successful middle class is inherently implausible, and provides the basis for Showalter's critique. However, when we looked outside the clinic, we found the opposite. Operationally-defined CFS was more common in lower socio-economic groups, but most of those affected did not use terms such as ME to describe their illness. There is thus no discrepancy between our epidemiological studies of CFS and Showalter's historical analysis of ME – we are describing different constructs and conditions, so it is not surprising that we reach different conclusions.[16]

Once a category is established, others may be allocated to it as a result of social identification or social labelling, rather than because they experience the organic condition. For example, the claim that autism is a genetic condition persists, yet the concept of an autistic spectrum disorder conceals the fact that some difficult children are diagnosed with an autistic disease when they do not show any measurable organic condition. The same dissonance between diagnostic category and biological evidence holds for Attention Deficit Hyperactivity Disorder, where the diagnosis is social while the rhetoric of explanation is biological. The recent work of Priscilla Alderson and Chris Goodey[17] challenges the 'reality' of autism, based on their observation in special schools. However, it seems that what they are challenging is the tendency of difficult children to be labelled autistic. They criticise the vagueness of the autistic category and question whether the condition itself has any unequivocal biological correlate. Clearly, both the range of autistic impairment and the lack of diagnostic clarity confuse the situation. If autism is an exaggeration of behaviour which is, in mild form, quite common in the general population – particularly the male population – then it is not surprising that there is difficulty in establishing who is neuro-typical and who is not. Here, too, there is an increasing tendency for those with the condition to celebrate and affirm difference, for example through the Inlv Internet forum. What for the dominant culture is a taken-for-granted connection between the difference, abnormality and pathology is commonly denied by the resistant culture of the disability movement.

Once a category is established, various social processes may take over. These include socialisation (the way that children are brought up to become members of a society, by learning the rules that govern it): this might explain why ambidextrous children become right-handed. That most tools are right-handed is not unimportant either as we see in the pleasure left-handed people take in special tool catalogues. The processes also include labelling, which is the way that society may allocate people to distinct categories as a consequence of particular differences. Medical diagnosis is still the classic form of labelling in contemporary culture. A third process is identification, which is when people voluntarily join a category for reasons which might be political, or social, or psychological: lesbians and gays 'coming out' is a good example, disabled people describing each other as 'crips' (from cripple) is another. Similarly people with AS (Asperger's Syndrome) speak of themselves as 'Aspies' as a positive alternative to 'Neurotypicals' (or 'NTs').[18]

Where the identification and classification of a particular phenomenon such as dyslexia is 'and/or', understanding the experience of people with these conditions requires returning to the both/and approach. So what is understood as the organic variety of dyslexia has to be located in specific social and cultural context. Particular methods of teaching English can reveal or compensate for dyslexia. For that matter, it seems that native Italian speakers who are dyslexic do not suffer the same high visibility as do native English speakers. The Italian written language is more phonetic, causing fewer problems for people with this learning difficulty.

Again, it is only through placing dyslexia in the wider context that we can understand a particular dimension of the experience of people with dyslexia. Recent research has showed that a very high proportion of petty criminals could be diagnosed as dyslexic. In a survey conducted in five London boroughs, 78 out of 150 offenders of average intellectual ability or better (as measured by IQ scores) showed indications of dyslexia or some similar learning difficulty. While there is no suggestion that dyslexia itself has direct behavioural consequences, it seems that some young working-class males with dyslexia feel alienated and scapegoated as school failures. They may suffer actual educational disadvantage, be unable to get jobs and drift into crime as a result of anomie and unemployment. That is, the social consequences of the way our society deals with (or fails to deal with) dyslexia are that some people with the condition in disadvantaged environments may have a higher likelihood of being criminally active.[19]

In these examples the condition and behaviours that people manifest have existed for long periods of time. Human beings have, almost certainly, always experienced same-sex attraction. The phenomenon can predate late twentieth century's diagnostic categories, hence the behaviours or syndrome now named as autism have a long history, and so on. However, the construction of the diagnostic categories of 'homosexual', 'autist', give rise to qualitative changes in the experience of individuals so categorised.

Labels/Badges

People who find themselves members of a group may find that membership has political or cultural ramifications which may be quite undesirable – such as the equation of left-handed people with evil, or prejudice against gays. It is particularly common for membership of a specific category to overwhelm all other characteristics, a process which sociologists speak of as 'identity spread'. Yet the negative label can also become a positive badge. The pink triangle of the Nazi concentration camp becomes the symbol of gay liberation and pride. As Michel Foucault wrote, 'One should not be a homosexual, but one who clings passionately to the idea of being gay.'[20]

Foucault is reminding us that we need to consider the ways in which selves are involved not only in being defined or labelled, but are also involved in resisting definition and in constructing categories and identities for themselves. And it is with respect to this point that we would argue the both/and approach offers significant advantages over other ways of making sense of the world. An adequate model of identity is one that can include a consideration of the ways that an individual sees him/herself, can recognise the ways in which others are involved in socially constructing the individual and, crucially, recognises the ways in which there are material constraints on individuals and social groups that may remove possibilities for action and/or thought. Whereas biologically based accounts or extreme social constructionist accounts may leave the individual isolated in terms of making sense of themselves, a both/and approach offers significant and grounded resources for identity formation. In addition, the both/and approach provides a critical social perspective upon which research and political engagement may be based.

How does this work? Our identities are formed through the assimilation and articulation of narratives, but these identities are not fixed and transcendental: they shift according to where we are, what knowledge is available to us and who we are with. So the construction of identity may change depending on where, or by whom, the description is being made. For example, at the doctor's surgery, a disabled person may be looking for a clinical response to the problems presented by their impairment. Yet, in the political arena, the demand for civil rights depends on a recognition of the socially generated disabling barriers which deny citizenship.

We tell ourselves the story of our self in a range of ways and base such stories on a range of resources. Either/or approaches present people with narratives based upon fixed circumstances to explain their situation, whether these narratives are based on genes or ideology. They also enable researchers to make sense of people by reference to such structuring themes. Yet they do not allow for the ways in which the stories people tell themselves relate to the stories which are told about them. In contrast, a both/and approach is sympathetic to the ways in which individuals see themselves and the ways in which we as social actors are involved in constructing the identity of others. There is a necessary circularity here, which is what Ian Hacking means by his term 'dynamic nominalism'.

Rather than seeing identities as being constructed purely externally, or seeing identities as solely a consequence of self-description, this approach allows us to understand the interplay of different descriptions from different sources. In particular, the both/and approach recognises that there is an interplay between the categorisation of people from above and the construction of identity by people from below,[21] and it allows us to see that this interplay may be structured by material conditions – such as bodily states or economic factors. There are parameters to our lives which limit our exercise of choice.

Of course, there is a 'pull to essentialism' in contemporary identity politics. Gays may vindicate their lifestyles in terms of the 'naturalness' of their genetic make-up. 'Aspies' may use biological categories in preference to those who would blame psychodynamic factors, damaged social roles, or mental illness for their behavioural quirks. People with ME wish to establish the physicality of their condition. We would not wish to exclude these possibilities. But we have to recognise that these processes inevitably take place in social and cultural contexts and cannot be separated from the structures of power associated with labelling.

The risk of basing identity in biology is that supposed biological difference has often been used to justify inferiority, whether in the case of black people, women, lesbians and gays, or disabled people. Thus Paul Gilroy reminds us that the tendency of slaves on American plantations to run away was rationalised by contemporary medical opinion as an illness known as 'drapetomania' or 'dyaesthesia Aetheopis', and Janet Sayers records the ways in which nineteenth-century women were excluded from higher education, and consequently public life, on obstetric and psychiatric grounds.[22] The use of contemporary genome research to reinforce existing prejudices is a variant of a familiar historical process.

In this chapter we have tried to show that biological and social and cultural processes weave together in complex ways to produce the phenomena which we experience. Just as the best versions of the biological story stress the dynamic processes in nature, the inextricable involvement of the environment with the expression of genes and the crucial role of development, so an adequate social science must acknowledge the bodily and ecological parameters within which humans operate.

Nothing is either natural, or social, as these concepts have been traditionally understood. Shallow reductionism – either explaining everything via evolved biology, or via social determinism, or the effects of language – leads to polarities which produce crude distortions of people's lives. Politics deploys stories for rhetorical effect, to establish critique or to mobilise constituencies for change. Individuals construct narrative identities in order to make sense of their emotions and in/abilities. Causes and categories are part of a symbolic grammar of being and becoming.

These are fine distinctions and multi-faceted relationships which are a long way from the *Just So* stories of evolutionary psychology or the single-gene determinism of the biological reductionists. Faced with the complexity of human lives, we could decide that there is no meaning and all we can do is play word games. Alternatively, in our desperate desire to explain the world it may be tempting to resort to the headline or the soundbite, or to substitute a precise genetic location for the messy nuance and variability of human behaviour. But, as Einstein said, 'Make everything as simple as possible. But not simpler.'

Notes and References

1 For example, Nikolas Rose, *Governing the Soul* (London, Routledge, 1989); Nicholas Fox, *Postmodernism, Sociology and Health* (Buckingham, Open University Press, 1993); Michele Barrett and Ann Phillips, *Destabilizing Theory* (Cambridge, Polity, 1992).
2 Judith Butler, *Gender Trouble* (New York, Routledge, 1990).
3 Harriet Bradley, *Fractured Identities* (Cambridge, Polity, 1996), p. 9.
4 Bruno Latour, *We Have Never Been Modern* (Cambridge, MA, Harvard University Press, 1993).
5 Jane Campbell and Michael Oliver, *Disability Politics* (London, Routledge, 1996), or other works by Oliver, summarises these arguments.
6 Ann Oakley, *Sex, Gender and Society* (London, Temple Smith, 1972).
7 Denise Riley, *Am I That Name?* (London, Macmillan, 1988).
8 Liz Crow, 'Including All of Our Lives: Renewing the Social Model of Disability', in Jenny Morris (ed.), *Encounters with Strangers* (London, Women's Press, 1996), p. 208.
9 Mary Douglas, *Purity and Danger* (Harmondsworth, Penguin, 1976).
10 Another stigmatised condition, epilepsy, can usually be explained by organic brain processes detectable by electroencephalography and treated by anti-convulsants. But one-fifth of cases do not display these biological signs and may be better treated by cognitive and behavioural therapies.
11 See, for example, Simon LeVay, 'A Difference in Hypothalamic Structure Between Heterosexual and Homosexual Men', *Science*, 253 (1991), pp. 1034–7. Women are usually ignored in these debates.
12 Steven Rose, *Lifelines* (London, Allen Lane, 1997).
13 Mary McIntosh, 'The Homosexual Role', *Social Problems*, 16 (1976), pp. 182–91; Michel Foucault, *The History of Sexuality* (Harmondsworth, Penguin, 1976).
14 Ian Hacking, 'Making Up People', in T. C. Helier et al. (ed.), *Reconstructing Individualism* (Stanford, Stanford University Press, 1986), p. 236.
15 A similar process may operate in the case of dyslexia.
16 Simon Wessely, 'Terminological illness', *London Review of Books* (17 September 1998), p. 4.
17 Priscilla Alderson and Chris Goodey, *Enabling Education* (London, Tufnell Press, 1998).
18 Judy Singer, ' "Why Can't You Be Normal For Once in Your Life?" From a "Problem with No Name" to the Emergence of a New Category of Difference', in Marian Corker and Sally French (ed.), *Disability Discourse* (Buckingham, Open University Press, 1999).

19 The 'Life as a disabled child' research project at the Universities of Leeds and Edinburgh has provided qualitative data about the ways that teenage boys with dyslexia may feel alienated and be labelled as troublesome before they have been identified as having a specific learning difficulty. After diagnosis and treatment, their social exclusion diminishes.

20 Michel Foucault, *Politics, Philosophy, Culture* (London, Routledge, 1990), p. xxiii.

21 For a fuller account of such 'labelling vectors' see Hacking, 'Making Up People'.

22 Paul Gilroy, *The Black Atlantic* (London, Verso, 1993); Janet Sayers, *Biological Politics* (London, Tavistock, 1982).

13

Social Causes and Natural Relations

Ted Benton

The Politics of Human Evolution

That immense superiority which the white race has won over the other races in the struggle for existence is due to Natural Selection, the key to all advance in culture, to all so-called history, as it is the key to the origin of species in the kingdoms of the living. That superiority will, without doubt, become more and more marked in the future, so that still fewer races of man will be able, as time advances, to contend with the white in the struggle for existence . . .

Melancholy as is the battle of the different races of man, much as we may sorrow at the fact that here also might rides at all points over right, a lofty consolation is still ours in that, on the whole, it is the more perfect, the nobler man that triumphs over his fellows . . .[1]

Now it is very remarkable that among people in a very low stage of civilization we find some approach to such a perfect social state. I have lived with communities of savages in south America and in the East, who have no laws or law courts but the public opinion of the village freely expressed. Each man scrupulously respects the rights of his fellows, and any infraction of those rights rarely or never takes place. In such a community, all are nearly equal. There are none of those wide distinctions, of education and ignorance, wealth and poverty, master and servant, which are the product of our civilization . . . Our mastery over the forces of nature has led to a rapid growth of population, and a vast accumulation of wealth; but these have brought with them such an amount of poverty and crime, and have fostered the growth of so much sordid feeling and so many fierce passions, that it may well be questioned, whether the mental and moral status of our population has not on the average been lowered.[2]

206

These quotations come from the writings of two of the most celebrated evolutionists of the nineteenth century. The first, Ernst Haeckel, identifies human cultural difference with biological race and uses the Darwinian concept of 'struggle for existence' to describe the violent conquest and race extermination carried out by the European imperialist powers as a regrettable, but inevitable instance of the progressive workings of natural selection. The second, Alfred Russel Wallace, characterises the social life of the indigenous peoples of the Amazon and south-east Asia as morally 'near perfect', by contrast with the moral and mental 'barbarism' of his own society. Wallace's view of the difference between savage and civilised modes of life turns Haeckel's racial hierarchy on its head. But it also differs from Haeckel's view in the way it understands mental and moral difference. For Wallace, it is the relative social equality and the normative power of 'public opinion' which explain the moral integration and mutual respect of the indigenous communities. By contrast, it is the accumulation of wealth, with its accompanying social divisions between rich and poor, which leads to mental and moral degeneration in the European society.

So, between two contemporaries, both committed 'Darwinian' evolutionists, there lies a huge gulf in the ways they think about racial and cultural difference. There were similar and related differences in the way the evolutionists of that time thought about difference between the sexes:

Woman seems to differ from man in mental disposition, chiefly in her greater tenderness and less selfishness . . . Man is the rival of other men; he delights in competition, and this leads to ambition which passes too easily into selfishness. These latter qualities seem to be his natural and unfortunate birthright. It is generally admitted that with woman the powers of intuition, of rapid perception, and perhaps of imitation, are more strongly marked than in man; but some, at least, of these faculties are characteristic of the lower races, and therefore of a past and lower state of civilisation.[3]

. . . [T]he position of woman in the not too distant future will be far higher and more important than any which has been claimed for or by her in the past. While she will be conceded full political and social rights on an equality with man, she will be placed in a position of responsibility and power which will render her his superior, since the future moral progress of the race will so largely depend upon her free choice in marriage. As time goes on and she acquires more and more economic independence, that alone will give her an effective choice which she has never had before.[4]

Darwin, the author of the first quotation, identifies current gender stereotypes with inherited sex differences ('natural birthright'), and links some female characteristics to those of the 'lower races'. By contrast, Wallace, the author of the second, presents an optimistic vision of the future social, economic and political emancipation of women and its beneficial consequences for 'the future moral progress of the race'. Of course, there are similarities too. Both men work with unquestioned gender stereotypes, but it is clear that Wallace leaves open far more possibilities for future transformation in the relations between the sexes than does Darwin. Moreover, Wallace's argument suggests that progressive social change will make a contribution to the continuing evolution of humanity.

So, for these Darwinian evolutionists, there was no fundamental disagreement about the origin of species, nor about the status of 'man' as an evolved primate species. But there were very big differences of view about the nature and fixity of observable differences between human groups: between 'savage' and 'civilised' races, men and women, and the social classes. These were, of course, related to wide differences in the moral and political outlooks of the evolutionists themselves. Darwin was a moderate liberal–progressive in outlook, Wallace a committed socialist and proto-feminist, whilst Haeckel moved, through his long career, from liberal–progressive to become the intellectual inspiration of a 'monist' movement which some have seen as precursor to the Nazi ideology.[5] However, it would be very misleading to see these differences in the way evolutionary ideas were used as just a matter of political preference. Underlying the contrasting attitudes illustrated in the above quotations were marked differences in interpretation of evolutionary theory itself and in ways of applying it to humans. These differences continue to resurface in the evolutionary debates of our own time, and, as in the nineteenth century, they continue to be inextricably bound up with differences of moral and political outlook.

Repeated attempts to extend evolutionary theory to explain human mental and social life have claimed the authority of 'science'.[6] Borrowing from the established status of Darwinian evolutionism has been seen as the way to overcome the factionalism and failings of the social sciences and put them on firm, scientific foundations. But such claims should never be taken at face value. It is possible to call up Darwin in support of just as many conflicting social and political doctrines today as it was in the nineteenth century. There are three basic reasons for this. One is that evolutionary biologists differ among themselves about the nature of the mechanisms of organic change, as illustrated by the contributions of several biologists to this volume. The second is that the theory is radically indeterminate in its implications for the way that any particular lineage might be expected to evolve. This will depend on the particular pool of genetic variation available for 'selection' in the course of its history, the unique constellation of shifting survival-relevant environmental conditions of life throughout that history, and the interactions of the successive generations of organisms with those conditions. In other words moving from general theoretical statements to particular applications of the theory is not (as in more abstract sciences) just a matter of logical inference, but of immensely detailed historical and ecological investigation. This is no less true of the lineage that led to modern humans.

So inferences from the theory of evolution by natural selection about evolved human nature depend on having the evidence to back up just such a complex narrative about our ancestors and their environments. The problem is that, though there is today much more evidence from palaeo-archaeology than there was in the days of Haeckel, Wallace and Darwin himself, it is still very limited, fragmentary and open to widely different rival interpretations. The third reason why inferences from evolutionary theory to explain human social life are contested is that human sociability is quite unlike that of other species. Human societies are complexly ordered, highly variable and subject to transformation on a historical rather than evolutionary time-scale. How far, and in what ways, evolutionary theory is relevant to these 'emergent' features is not a straightforward scientific question, but is subject to intense theoretical and philosophical debate. It follows that any attempt to 'close' these open questions is bound to involve a lot of speculation and special pleading. The space between what is known and what has to be assumed is particularly liable to be filled with unquestioned political and moral prejudices.

Darwin is credited with the invention of the concept of natural selection, but he continued to believe that other mechanisms were also at work. Many evolutionists (and this was particularly true in Germany, where there was already a well-established pre-Darwinian tradition in biology) made no clear distinction between natural selection (operating through differential reproduction of inherited traits) and a broader notion of organic change occurring under environmental influences. If, as many did, one included social and economic conditions among these environmental influences, then evolutionism could easily be harnessed to the cause of progressive social change. However, this confusion became more difficult towards the end of the nineteenth century, as the idea of the inheritance of characteristics acquired during an individual's lifetime became discredited. From then on, evolutionary 'progress' in the human case came to be seen as largely dependent on differential reproductive rates of more or less 'fit' specimens of the race. This was the key issue which made the debate about 'nature' versus 'nurture' so important: for those who were concerned about the impact of social conditions on human wellbeing, it became necessary to set limits to the reach of explanations in terms of biological evolution.

For many, this entailed setting humans apart from other animals. Wallace, for example, though he believed in natural selection as the key mechanism of organic change, could not accept that it was adequate to explain distinctively human traits such as humour, aesthetic appreciation and ability in higher mathematics. This led him in later life into spiritualism and to a belief that supernatural forces were at work alongside natural selection in human evolution. But it was not necessary to take such an extreme route. Given that evolution by natural selection was a very slow process, it had to be assumed that the very different human societies of the historical past were made by humans biologically little or no different from those of the day. Evidence was there from travellers' tales, and from the increasingly powerful drive to colonisation and empire, not only that 'savages' had radically different forms of social life, but also that they could be taught the ways of 'civilisation'. Humans were malleable, born unformed and open to being shaped mentally and morally by the social environment they entered into.

It was this environmental view that prevailed in the great struggles of the nineteenth and twentieth centuries to overcome poverty, ignorance and material insecurity, to extend the rights of propertyless wage workers, to enhance the social, political and economic status of women, to emancipate racial and ethnic minorities, and to dismantle empire. Alongside these struggles, and to some extent intertwined with them, there were established academic disciplines – anthropology, sociology, political science, social policy, especially – which took the social processes at issue in these struggles as their special subject matter. Defending the right of these disciplines to independent existence meant showing that biological evolution had its limits. It might be able to explain human *origins* (though some continued to doubt even that), but once humans had evolved their distinctive character, a quite different sort of understanding was required to make sense of their *subsequent* modes of social and cultural life. For some of the founders of this cluster of new social sciences, this difference flowed from the acquisition of language and meaning, whilst for others it was more a matter of the emergence of a new order of 'social facts' as a consequence of the association of humans in shared forms of life. Either way, social and cultural processes could be identified as subject matters in their own right, not reducible to the much slower processes of organic change which were the proper subject matter of evolutionary biology. This strategy for distancing evolutionary biology from the concerns of the social sciences fitted neatly into a wider division of labour between the sciences of nature and the 'human', 'cultural' or social sciences which has been maintained into our own time.

This division was, of course, always contested. The cultural authority of the idea of 'evolution', particularly with its popularised connotation of 'progress', proved irresistible for many of the early sociologists, and it was common to reconstruct the variety of past human societies into an evolutionary series, marked by progressive development.[7] Generally, however, these social evolutionary ideas relied on specifically social explanations of the advance from one evolutionary stage to the next. Set against these were the continuing advocates of the 'nature' side in the nature/nurture controversy. They have tended to be on the side of resistance to the emancipatory struggles of the last century and a half. Their aim has been to show that the profound inequalities of social opportunity, recognition and life chances associated with class division, racial and gender domination and social hierarchy are 'natural', largely the result of unalterable inherited differences. In their most virulent forms they have provided intellectual comfort for projects of eugenic racial purification and outright genocide. However, the legacy of cultural horror left in Europe by the Holocaust has put all would-be biological determinisms sharply on the defensive.

A Sociological Mutation

The last two decades have seen a marked downturn in the influence of the traditional movements of the progressive left. This has coincided with the re-emergence, as popularised secular religion, of new forms of genetic determinist ideology. Sociology, as Hilary Rose indicates, has long been hostile to biological determinism, thus the embrace of EP by the distinguished sociological theorist and peer, W. G. Runciman, cannot go unremarked. In *The Social Animal* and in a series of related articles[8] Runciman has advocated a 'Darwinian' approach to sociology. As is common with social evolutionists, he slides back and forth between direct evolutionary explanation of social life, and a more modest 'Darwinism-by-analogy'. Although Runciman seems to owe more to earlier forms of social Darwinism than to EP, he endorses the claims of the latter, both in its interpretation of evolutionary theory, and in its attack on the 'Standard Social Science Model' (about which more later):

'There can be no question that this capacity [for culture – TB] is a product of natural selection to be explained by its function in augmenting reproductive potential in an environment which over very many generations favoured the carriers of the relevant genes.'[9]

And, a little later:

researchers still attracted to the paradigms of either Skinnerian psychology[10] on the one hand or Durkheimian anthropology on the other may simply not be aware of the implications of the recent advances in evolutionary, cognitive and developmental psychology, which have radically undermined the 'Standard Social Science Model'[11]

One can only wonder about the reasons why a sociologist with so much experience of the rich analytical traditions available in his own discipline might wish to convey the choice as merely between Skinner and Durkheim. One possible explanation is the long legacy of envy on the part of some social scientists for the degree of apparent unity and social prestige enjoyed by the natural sciences: the paradigm of Darwinian evolutionism is invariably popularised as the authoritative voice of 'Science', as if this settled everything.

Contemporary anthropological accounts of kinship, from for example Marshall Sahlins to Marilyn Strathern, point out that genetic and social constructions of kin have by no means a one-on-one relationship.[12] Genetic kin selection as proposed by sociobiologist William Hamilton therefore cannot explain patterns of behaviour, even in those societies in which kinship is a central organising institution. Yet Runciman reads the anthropological evidence very curiously, arguing that what is most important is that such societies are primarily organised around kinship, whether this is genetically or socially constructed. It is hard to make sense of this argument, unless Runciman sees the symbolic constructs of kinship as in some sense a 'false consciousness' whereby predispositions to favour genetic kin are hijacked by the symbolic order to produce quite different outcomes, not least behaviours.

Although Runciman thinks that a significant amount of human behaviour may be 'parsimoniously explained by hypotheses derived directly from natural selection', he accepts that there are emergent characteristics 'whose explanation is irreducible to the biological, let alone the chemical, level'. Like most other sociologists, Runciman recognises that the interpersonal understanding of meaning is an irreducible constituent of social life. It follows that social explanation has to make reference to these meanings, to institutions, to the roles, power relations and so on, which hold between persons as social agents, not merely as bearers of a certain genetic inheritance. We are thus 'no longer in a world of instinctive cooperation or conflict, as at the biological level . . . but of agents' power to affect each other's behaviour because of their institutional position'.[13]

Having conceded that evolution by natural selection is unable to provide explanations of such forms of social action, Runciman's fallback position is that some process *analogous* to natural selection is needed to explain social and historical change and diversity. Explanation in terms of 'descent with modification' is the only sort of explanation that could be scientifically acceptable. To meet this requirement, Runciman needs to be able to identify 'units' of social life analogous to genes as units of biological evolution. For Runciman, this means that they have to be capable of replication, at varying rates depending on the operation of environmental (in this case social environmental) selective pressures, but also with occasional 'copying errors', analogous to genetic mutations. (Also see Mary Midgley's chapter.)

So far, Runciman's approach is close to the 'meme' view of culture, but he (again, like most other sociologists) does not accept that social life is just 'culture'. In fact, he distinguishes three types of human behaviour. 'Evoked' behaviour is instinctual, and presumably explicable directly in terms of natural selection. 'Acquired' behaviour is adopted by individuals on the basis of learning or imitation and is presumably equivalent to culture. The third sort of behaviour is 'imposed' and is 'defined by institutional rules which are not of the agent's making'. It is this type of behaviour, according to Runciman, that is of particular interest to sociology. The units (analogous to genes) for sociological explanation are what he calls 'practices'. These are defined as 'units of reciprocal behaviour informed by mutual recognition of shared intentions and beliefs'. These units are the basic components of social roles, institutions and so on, making up society, and human agents maintain society by replicating practices. As he puts it, we 'are machines for replicating the practices which define those roles and the groups, communities, institutions and societies constituted by them'. (Note the echo of Dawkins's language of humans as lumbering robots replicating our selfish genes.) So long as the human machines keep replicating the practices, social stability is maintained. However, sometimes copying errors occur – presumably through innovative or deviant performances of existing practices – and if the social environment favours the replication of the new variant it will spread throughout the society. This is the Darwinian 'descent with modification' model for explaining social and historical change.

As a research programme for sociology, there are three main reasons why this is unworkable. First, the analogy between practices and genes doesn't stand up to serious examination. Second, the whole idea of practices as basic 'units' or building blocks of society, and of societal change is inconsistent with basic common knowledge about social life, and, indeed, with many things which Runciman himself says. Third, the view of humans as machines for replicating practices involves a fundamental misunderstanding of individual human participation in social life. This further undermines the analogy Runciman claims between his approach to sociology and Darwinian evolution.

The analogy between genes and practices breaks down continually. First, how are unit practices identified? Processes of social interaction are continuous, so where do we draw boundaries – start and end of conversations? Initiation and completion of a work task? Start and end of sentences, or arguments, or phrases? Exchanges of facial expressions? Crossing the road, stepping out to cross the road, signalling that we are about to? . . . Or? In classical Mendelian genetics, there are distinguishable phenotypic characters that were subsequently linked to particular alleles. This is how 'genes' as units could be indirectly inferred prior to their microscopic identification and ultimately biochemical analysis. But unit practices cannot be tied down in any of these ways. There is no basis for a distinction between genotype and phenotype – practices are not underlying entities that produce societies by way of embryonic development in the way genes contribute to the development of organisms. In the absence of such connections the criteria by which the boundaries of practices are set are contextually defined. 'I went shopping' will count as the description of a unit practice if all that is needed is an account of where I was when someone failed to find me at home. But if I slipped on the pavement as I crossed the road on the way to the shop and caused a crash, a more fine-grained discrimination of unit actions would be needed to find out if I had slipped on a loose kerb, warned oncoming motorists of my intentions, and so on.

But the key idea of practices as the units out of which societies are built is itself flawed. Runciman defines unit practices as involving 'shared intentions and beliefs'. But beliefs and intentions themselves cannot be identified independently of their logical connection with wider patterns of understanding. For a motorist to understand my intention when I raise an arm and step on to a pedestrian crossing presupposes a whole complex of interconnected knowledges, skills and meanings: the crossing signs themselves, the rules governing priorities in this society in this situation, the reading of gestures and expressions, general expectations about people's need to cross roads and so on.

Virtually all of this is tacit, but it is there none the less, and the consequences when it goes wrong can be both manifest and severe.

So, the point of this is that 'unit' practices can only exist in virtue of broader socio-cultural frameworks that provide people with the means to identify the beliefs, meanings and actions which define their practices as what they are. Each practice is located within assumptions about both culture and society. This is clearly implied by Runciman's own definition of 'imposed' behaviour, according to which action is governed by institutional rules 'not of the agent's making'. Here it is not just a matter of cultural context being needed to give identity to a practice, but of wider structures of institutional power which shape and direct the activity of people subject to them. To see this is to admit that 'society' is not just a complex built up out of individual practices as its 'building blocks'. Atomistic accounts of society like this serve to obscure the key insights given to us by sociology – insights into the many complex and more or less subtle ways in which society and our place in it shape us and significantly affect our chances for life, liberty and happiness.

Finally, what are we to make of Runciman's account of human agents as 'machines for replicating practices'? If the analogy worked it would settle the problem of how to distinguish a replication from a 'mutation'. Individual performances of 'the same' practice would be standardised in the way that, for example, different copies of a CD or different playings of the same CD would generate the same sequences of musical sounds. But this is not how human practices are 'replicated'. Think of different performances of the same musical score by the same quartet at different concerts or by different groups of musicians on different occasions. The distinction between mechanical reproduction and the reflexivity and intentionality involved in the social rule-following of the musicians is central both to living and understanding social life.[14]

Runciman needs the machine-view of social agency partly to shore up his analogy between practices and genes, but also because he wants to make a point about how social and historical change occur. He wants to downplay the role of human agents, introducing change to realise human purposes. Instead he wants to emphasise the way societies change as a result of impersonal forces, 'environments', which favour or disfavour the 'replication' of mutant practices on the part of replicating 'machines'. The parallel with Darwin's problem in explaining evolutionary change is a very interesting one. Darwin was deeply impressed with the 'fit' between organisms and their ecological niche. This appearance of design had provided a powerful argument for those who believed that life forms owed their creation to a super-natural designer: that intentional agency was at work. Darwin's hypothesis of natural selection undermined this argument by showing how the appearance of design could be produced by natural causes without need for a purposive agent.

Runciman applies this model of explanation to human societies and their historical change. But in this case changes are, at least in part, brought about by intentional agents, individually or collectively. Darwin's mechanism of natural selection assumes that mutations are random with respect to the selective pressures which affect their chances of replication. In the case of the activities which lead to social change – the struggle of women for the vote, the government privatisation of publicly owned industries, the development of new technologies, the invention of new sociological theories and so on – human agents act intentionally to produce anticipated outcomes: they are not 'blind watchmakers'. The problem which Darwin solved was how to explain the appearance of design without a designer. Runciman seems to have failed to notice that sociologists have a different problem. Societies, unlike species, do have designers. What sociologists have to explain is why the designs so often don't work! So there is a place for unintended consequences, contingency and impersonal forces in historical explanation: but these are the results of the institutional structures, power relations, social conflicts, mistaken assumptions, betrayals, communicative failures, and so on, of intentional agents, not of 'machines for the replication of practices'.

Lastly, when Runciman does offer empirical illustrations of the utility of his version of Social Darwinism it is difficult to avoid the impression that his narratives get whatever plausibility they have from the primary research of orthodox social scientists, whose work Runciman simply presses into the service of his own terminology: old wine, new bottles.

Evolutionary Psychology: A Contradictory Inheritance

The work of Runciman's mentors, the evolutionary psychologists, suffers from a similar slippage back and forth between outright genetic determinism and a much more modest and cautious programmatic rhetoric. But this slippage is not just a matter of inconsistency; it flows from basic tensions in the whole approach. What EP shares with previous Social Darwinisms is its mission to undermine the foundations of the existing social science disciplines and to bring them within the fold of a particular version of evolutionary biology. As we have seen, this has important moral and political implications. Meanwhile EP persistently overrides its own methodological cautions and lapses into full-blown biological determinism. It does so in part because of the logic of its own version of evolutionary theory and partly because it is in the grip of an unquestioned reductionist view of the relation between social and psychological processes.

One of the many problematic features of the version of evolutionary theory employed by EP is that it understands 'adaptation' as a one-way process through which populations of organisms increasingly meet survival requirements posed by an implacable independent 'environment'. In their account of hominid evolution the evolutionary psychologists have to build a lot of speculation on to the very fragmentary sources of evidence available from the fossil record. They supplement this with reference to ('naturally', very selective!) appeals to the social lives of contemporary hunting-and-gathering peoples, as if this self-evidently tells us about the selective pressures at work on our evolutionary ancestors. This is, of course, methodologically unacceptable: not only do contemporary hunter-gatherers inhabit many different environments quite unlike the African savannah, also they are not time-warp relics of our hominid ancestry, but evolved humans.

The view of natural selection and its adaptive outcomes which the evolutionary psychologists employ leads them to explain the evolution of distinctively human psychological mechanisms in terms of their entry into the 'cognitive niche'.[15] The development by hominids of minds/brains as complex modular systems of information processors and behaviour-generators enabled them to outwit competitors by devising novel, goal-oriented courses of action on a time-scale far shorter than could be achieved by natural selection. This does acknowledge a role for learning and variability in behaviour. However, for EP, this is learning shaped and to a degree preprogrammed to generate behaviour likely to be adaptive in the putative savannah environment. On this view, selective pressures would have led to genetic 'hard-wiring' of basic 'intuitive theories' about objects, animals and other people. This, in turn, would have conferred a selective advantage by making the learning requirements on individuals of each new generation so much less demanding.

The obvious difficulty with this account is that there is an unresolved contradiction at the heart of it. On the one hand, it is claimed that the selective advantage conferred by brain development is the way it opens up the possibility of flexible and innovative responses to environmental challenges. On the other, it is also claimed that selective pressures would lead to a progressively 'hard-wired' set of learning dispositions and 'behaviour-generators'. The unresolved tension between these different elements in the EP story is one major cause of their confused and contradictory zigzags between theoretical modesty and extreme biological determinism. Steven Pinker, for example, after endorsing Tooby and Cosmides's distinction between universal psychological mechanisms and culturally variable behaviour, cannot resist the temptation to provide us with accounts of happiness and virtue, the full complement of human emotions, supposedly universal preferences in the visual arts (for scenes reminiscent of our past on the ancestral savannah), love of music, and sense of humour, all in terms of the survival requirements of the African savannah!

Not only is this story of progressively hard-wired adaptation to a 'cognitive niche' internally inconsistent, it also squarely contradicts such evidence as is available. If, indeed, the selective advantage that brain development conferred was an ability to respond flexibly, creatively and quickly to environmental challenges, then this would suggest an evolutionary trend towards more open-ended learning and regulation of behaviour and away from 'hard-wiring'. The evolution of tool use and manual dexterity, together with language and conventionally regulated social co-operation, suggests a direction of development not well described as 'adaptation' to a particular 'niche'. On the contrary, these features taken together suggest the emergence of populations capable of making themselves increasingly independent of particular 'niches', of moving 'across' and 'between' the niches occupied by other species and by none:

> Where diggers had needed heavy nails, now there were stone picks, cats no longer had the monopoly of sharp claws, spears mimicked horns, porcupine quills or canine teeth and so on. Here, for the first time, was an animal that was learning a multiplicity of roles via the invention of technology. An increasing number of animals now had a new competitor that would encroach on at least a part of their former niche. In some cases – perhaps some of the scavengers – the overlap became so great that the hominids took over.[16]

Indeed, the degree of flexible interaction with features of the environment implicit in these developments makes the concept of 'niche' itself too hopelessly fixed and static to illuminate the processes at work. Instead, the process would be better characterised not as one of 'adaptation' to a fixed 'niche' but of the evolution of a quite new order of flexible adaptability, enabling both survival in the face of rapidly changing environments and successful exploitation of an indefinite variety of new ones. It is this which made possible the spread of hominids into a vast diversity of ecosystems across the 'old world' as long as 1.5 million years ago. Social groupings were themselves doing the 'selecting', and at the same time redesigning their social relations and material cultures to exploit quite new local environments on a time-scale quite inconsistent with slow-moving natural selection. The key point here is that 'flexible adaptability' on this scale implies a very large gap between inherited psychological powers and dispositions, on the one hand, and actual patterns of behaviour, on the other (what Steven Rose refers to as the difference between distal and proximal causes). In short, a very much larger space is opened up for psychological and social scientific investigation of the *non-genetic* sources of human activity.

Genes, Brains and Societies

The evolutionary psychologists' inability, in the end, to resist the drift into genetic determinism has another important source. This is their view of the relationships between social, cultural and psychological processes. Steven Pinker puts it this way: 'The geneticist Theodosius Dobzhansky famously wrote that nothing in biology makes sense except in the light of evolution. We can add that nothing in culture makes sense except in the light of psychology. Evolution created psychology, and that is how it explains culture.'[17]

The editors of *The Adapted Mind*[18] echo this:

Culture is not causeless and disembodied. It is generated in rich and intricate ways by information-processing mechanisms situated in human minds. These mechanisms are, in turn, the elaborately sculpted product of the evolutionary process. Therefore, to understand the relationship between biology and culture one must first understand the architecture of our evolved psychology.

Both these accounts assume a one-way causal chain from the evolution of inherited 'developmental programmes', which install information-processing mechanisms in human minds, through to the 'generation' of 'culture'. The evolutionary psychologists defend this scheme for linking evolutionary biology, psychology and culture against the 'Standard Social Science Model'. According to them, this is the model of explanation in the social sciences (including most of psychology) that is common to the otherwise rival traditions in these disciplines. This model of explanation, they claim, has prevented the social sciences from linking up with other sciences, including evolutionary biology, and this is why they have failed to achieve scientific status.

According to Tooby and Cosmides, the SSSM insists on the autonomy of culture. It may admit that humans do have a common inheritance of basic psychological drives and general learning mechanisms, but it insists that environmental influences, mainly the local culture, are responsible for the actual 'realised' behaviour patterns of individuals and whole societies. Individuals are represented as mere passive recipients of this socialisation process. According to the EP stereotype, this is how the social sciences account for the great diversity of human cultures.

Evolutionary psychologists agree that human cultures are very variable, and that learning plays a big part in acquisition of language and other cultural attributes. They also agree on the biological unity of the human species. But their alternative 'Integrated Causal Model' (ICM) holds that humans inherit complex, specialised and content-specific psychological mechanisms which process information from the world in such a way as to generate richly variable culture and behaviour. Proponents of the SSSM are held to be mistaken in thinking that such specialised and content-specific mechanisms constrain culture – on the contrary, humans can only perform the complex learning tasks they do because they have this inherited mental equipment. So human flexibility and cultural variability are not in doubt:

> [W]e know in advance that the human psychological system is immensely flexible as to outcome: Everything that every individual has ever done in all of human history and prehistory establishes the minimum boundary of the possible. The maximum, if any, is completely unknown. Given the fact that we are almost entirely ignorant of the computational specifics of the hundreds or thousands of mechanisms that comprise the human mind, it is far beyond the competence of anyone living to say what are and are not achievable outcomes for human beings.[19]

It would be hard to find a social scientist who would dissent from this very modest statement. So, the opposition between the SSSM and the preferred ICM seems to come down to just two issues, one merely a matter of degree. First, how much inherited 'architecture' there is in the human mind. Second, whether diversity and flexibility should be explained in terms of the way this 'architecture' processes environmental inputs or in terms of autonomously acting 'emergent' sociocultural processes.

But things are not as simple as this. First, the SSSM is largely a product of the evolutionary psychologists' own information-processing mechanisms. Hilary Rose effectively exposes this as the ignorant caricature it is. Evolutionary psychologists are trying on one of the oldest tricks in the book: make your own position seem eminently sensible by portraying the only alternative to it as a daft extremism. But this does not mean that there are no differences between EP and its opponents. The key point of difference is not over the 'blank sheet' view of the mind, but whether sociocultural processes are understood as independent or reducible to inherited psychological mechanisms. However, the EP method of 'reverse engineering' implicitly contradicts such simple reducibility. According to Pinker,

> For ninety-nine percent of human existence, people lived as foragers in small nomadic bands. Our brains are adapted to that long-vanished way of life, not to brand-new agricultural and industrial civilizations. They are not wired to cope with anonymous crowds, schooling, written language, government, police, courts, armies, modern medicine, formal social institutions, high technology, and other newcomers to the human experience.[20]

And the editors of *The Adapted Mind*:

> What we think of as all of human history – from, say, the rise of the Shang, Minoan, Egyptian, Indian, and Sumerian civilizations – and everything we take for granted as normal parts of life – agriculture, pastoralism, governments, police, sanitation, medical care, education, armies, transportation, and so on – are all the novel products of the last few thousand years.[21]

Since, on their own account, all these massive changes and differentiations in human social life have taken place on the basis of a universal and unchanging set of psychological mechanisms, it follows that other causal mechanisms, irreducible to this common inheritance, must have been at work. Even if we were, for the sake of argument, to accept the EP model of diversity through psychological processing of diverse environmental inputs, then this implies that the environmental 'inputs' make all the difference. And, of course, the 'environments' concerned here are the sociocultural ones they list – education, formal institutions, governments, police, new technologies and so on. So sociocultural processes do, after all, constitute a causal order in their own right, capable of shaping the behavioural patterns of individuals, albeit often by way of their inner psychological processing. Still more, they distribute individuals to positions in the institutional structures of society in ways which may be decisive for their opportunities for health, happiness and life itself.

As we have seen, Tooby and Cosmides insist that universality at the level of inherited psychological mechanisms does not constrain, but rather enables diversity and flexibility at the level of behavioural and cultural outcomes – to the extent that the outer boundaries of that flexibility are 'completely unknown'. But this caution about the possibility of human universal behaviours is not shared by other EP theorists, notably Pinker, who unproblematically insists on the universality of 'prestige and status, inequality of power and wealth, property, inheritance, reciprocity, punishment, sexual modesty, sexual regulation, sexual jealousy, a male preference for young women as partners, a division of labour by sex'.[22] So much for the claim that outer boundaries of human flexibility are 'completely unknown'! So much, also, for the view that complex, inherited psychological processing enhances rather than constrains human flexibility! The pressure to justify value-driven political conclusions overrides repeated programmatic rhetorics of modesty and caution.

An Alternative Integration

Despite EP's ideologues it is surely right to insist on the value of making links across disciplinary boundaries – including links between the natural and the social sciences. There is an urgent practical as well as intellectual case for this, as others in this book have argued. We inhabit a world in which the globalisation of markets and the associated increase in capital flows across national boundaries pose profound practical issues. Ecological devastation of the conditions of life for many of the poorest people on earth, the emergence of global processes of ecological degradation, the disintegration of local forms of democracy and the regulation of resource use, the widening democratic deficit, the growth of 'social exclusion' even in the richest countries, the uncontrolled spread of the means of violence and the upsurge of ethnic violence and genocide are all pressing issues which demand intelligent and practical responses. Coherent and imaginative interdisciplinary thinking which refuses to be limited by the inherited academic division of labour is urgently needed.

Over recent decades new approaches in the social sciences have begun to rise to this challenge, partly as a result of the cultural work of social movements committed to social and economic justice, peace, women's rights, racial equality and ecological protection. Where EP seems to recognise human universals in the shape of psychological mechanisms, or, alternatively, in hierarchy, sexual jealousy, the sexual double-standard, economic inequality and so on, the alternative view sees human solidarity stemming from shared features of evolved human life. As embodied beings, we share a vulnerability to disease, mortality, needs for food, water and shelter, sexuality, reproduction and our need to care for the young over a long period. We share both needs and capacities to co-operate, and we do this through forms of symbolic and normative communication which differ widely from place to place, and change through historical time.

On this view society is not just the outcome of mental processing on the part of the individual people who make it up: society is not even just made up of people. Through social relations 'society' structures the activities and abilities of people which bind them and their activities together. Such structures and their associated roles also bind people together with tracts of land, agricultural ecosystems, natural resources, tools, buildings, material infrastructures, means of communication and transport, not to mention populations of millions of wild and domesticated animals and plants. It is relationships such as these, not just inherited psychological dispositions, that both set horizons of future possibility, and impose constraints, for particular versions of human civilisation. Integration and co-operation between the sciences – not evolutionary reductionism – is the only intelligent way forward.

Notes and References

1 Ernst Haeckel, *The Pedigree of Man and Other Essays*, trans. Edward Aveling (London, Freethought, 1883), p. 85.

2 Alfred Russel Wallace, *The Malay Archipelago* (New York, Dover, 1962 [London, Macmillan, 1869]), pp. 456–7.

3 Charles Darwin, *The Descent of Man, and Selection in Relation to Sex*, 2nd edn (London, John Murray, 1874), pp. 563–4.

4 Alfred Russel Wallace, *Social Environment and Moral Progress* (London, Cassell, 1913), pp. 147–8.

5 Daniel Gasman, *The Scientific Origins of National Socialism* (London, Macdonald, 1971).

6 See, for example, Helena Cronin, 'It's Only Natural', *Red Pepper* (August 1997), p. 21.

7 See Peter Dickens, *Social Darwinism* (Milton Keynes, Open University, 1999).

8 See, in particular, W. G. Runciman, *The Social Animal* (London, HarperCollins, 1998) and 'The Selectionist Paradigm and Its Implications for Sociology', *Sociology*, 32 (February 1998), pp. 163–88.

9 'The Selectionist Paradigm', p. 166.

10 B. F. Skinner was the pioneer of behaviourist psychology, with its tabula rasa view of the young. Runciman's attack is a pure straw-person strategy as this has been an obsolete paradigm within psychology for almost three decades.

11 'The Selectionist Paradigm', p. 177.

12 For example Marshall Sahlins, *The Uses and Abuses of Biology* (London, Tavistock, 1977).

13 'The Selectionist Paradigm', p. 175.

14 See Peter Winch, *The Idea of a Social Science* (London, Routledge & Kegan Paul, 1958), and subsequent controversy, collected in B. Wilson (ed.), *Rationality* (Oxford, Blackwell, 1974).

15 John Tooby and Irvine De Vore, 'The Reconstruction of Hominid Evolution Through Strategic Modelling', in W. G. Kinzey (ed.), *The Evolution of Human Behavior* (Albany, NY, SUNY, 1987).

16 Jonathan Kingdon, as quoted in Chris Stringer and Robin McKie, *African Exodus* (London, Random House, 1996), p. 25.

17 Steven Pinker, *How the Mind Works* (Harmondsworth, Allen Lane, 1977), p. 210.

18 Leda Cosmides, John Tooby and Jerome H. Barkow, in Jerome H. Barkow, Leda Cosmides and John Tooby (ed.), *The Adapted Mind: Evolutionary Psychology and the Generation of Culture* (Oxford, Oxford University Press, 1992), p. 3.

19 Tooby and Cosmides in ibid., p. 40.

20 Pinker, *How the Mind Works*, p. 42.

21 Cosmides, Tooby and Barkow in *The Adapted Mind*, p. 5.

22 Pinker, *How the Mind Works*, p. 427.

14

Evolving Skills

Tim Ingold

How We Walk

Human beings are restless creatures. They are always moving about. But far from being haphazard these movements are, for the most part, highly controlled. In many cases this control is achieved through long and frequent practice; we may then speak of it as a skill. Now it is evident that people raised in different environments, and following different ways of life, also possess a range of different skills. As an anthropologist I am particularly concerned to understand the nature of these differences. It has long been conventional to attribute them to something called 'culture'. Whether culture is genuinely unique to human beings, or present – albeit in rudimentary forms – in non-human species, has been much debated. All agree, however, on two points. First, humans rely on culturally acquired skills to an extent unparalleled elsewhere in the animal kingdom. Second, whatever biological differences may exist among human beings, they are irrelevant so far as their acquisition of culture is concerned. Or, to put it another way, every creature born of man and woman should, in principle, be capable of acquiring the skills appropriate to any form of cultural life. 'One of the most significant facts about us', wrote the anthropologist Clifford Geertz, 'may finally be that we all begin with the natural equipment to live a thousand kinds of life but end in the end having lived only one.'[1]

Consider, for example, that most widespread of human movement skills: the ability to walk on two feet. Every human newborn, barring accident and handicap, has the potential to develop full bipedality. In that sense we are inclined to suppose that walking is an innate capacity, one for which – as Geertz would say – humans are naturally equipped. It is part of our biological make-up, given from the start rather than culturally acquired. Yet we also know that people in different societies

225

are brought up to walk in very different ways. One of the first to recognise the significance of this fact, as an index of cultural variation, was the French ethnologist Marcel Mauss, who set out to classify the extraordinarily diverse postures and gestures adopted by people around the world in their most ordinary activities, whether at rest (sleeping, squatting, sitting, standing) or in movement (walking, running, jumping, climbing, swimming). Thus:

> The *habitus* of the body being upright while walking, breathing, rhythm of the walk, swinging the fists, the elbows, progression with the trunk in advance of the body or by advancing either side of the body alternately (we have got accustomed to moving all the body forward at once). Feet turned in or out. Extension of the leg. We laugh at the 'goose-step'. It is the way the German army can obtain the maximum extension of the leg, given in particular that all Northerners, high on their legs, like to take as long steps as possible. In the absence of these exercises, we Frenchmen remain more or less knock-kneed . . .

For the adult in any society, Mauss concluded, walking is an acquired technique. There is no 'natural way' of going about it.[2]

How are we to reconcile these two views of walking: as innate capacity and acquired skill? One commonly proposed solution is to argue that while humans are naturally endowed with the anatomy that makes bipedal locomotion a practical possibility, and the behavioural propensity or 'instinct' to put it to effect, precise directions about *how* to walk are passed on from generation to generation as part of a cultural tradition. This tradition includes rules and representations laying down standards of propriety, perhaps specific to age and gender, that walkers are enjoined to follow, and in terms of which their performance is evaluated and interpreted. Thus while the capacity to walk is a biological universal, particular ways of walking are expressive of social values. Would it not suffice, then, to combine the biology of human nature with the sociology of cultural difference to produce a complete 'biosocial' account of the ways people walk?

Mauss thought it would not. For the link between human nature and culture can only be established by way of a third term, namely what is called the 'human mind'. Any account of the relation between biological and sociological dimensions of human existence must leave room, said Mauss, for the 'psychological mediator'.[3] More recent contributions to culture theory, drawing much of their inspiration from developments in cognitive science, have gone on to propose that if rules and representations for the generation of culturally appropriate behaviour are to be transmitted from one mind to another, across the generations, then certain devices must already be in place. These must enable the novice to 'decode' the input of sensory data drawn from observations of the behaviour of experienced practitioners and thereby to reconstruct these rules and representations inside his or her own head. Or as Roy D'Andrade, one of the pioneers of cognitive anthropology, claims, the transmission of specific cultural content, in the form of *programmes*, depends upon the functioning of universal cognitive capacities, or *processors*.[4]

Thus in language learning it is supposed that the child's acquisition of his or her mother tongue depends on the pre-existence, in the mind, of an innate language acquisition device (LAD), able to process the input of speech sounds so as to establish a system of grammatical and syntactical rules for the production of well-formed and comprehensible utterances. (But see Annette Karmiloff-Smith's chapter.) Likewise, there ought to exist a 'walking acquisition device' – a cognitive module dedicated to the construction of a culturally specific programme for bipedal locomotion from observations of other people's movements. Such a device should, in principle, be universal to the human mind. To complete our picture of the walking, talking human being we therefore have to put together three things: (i) the human body with its built-in anatomical structures and capacities of movement (limbs for walking, vocal tract for speech); (ii) the human mind with its hard-wired, computational architecture of processing mechanisms; and (iii) the assemblage of culturally specific representations or programmes whose transmission across the generations these mechanisms make possible.

I refer to this idea of the human being as the sum of three complementary parts, namely body, mind and culture, as the *complementarity thesis*. It is backed by a formidable intellectual alliance between the theoretical paradigms of neo-Darwinism in biology, cognitive science in psychology and culture theory in anthropology. Far from advocating this alliance, I shall argue that it is dangerously misconceived. Before doing so, however, I should explain how its constituent parts fit together, beginning with biology.

The Complementarity Thesis

1 Evolutionary biology. The central claim of Darwinian biology is that human beings, along with creatures of every other kind, have evolved through a process of variation under natural selection. This claim, however, rests on the critical assumption that the growth and maturation of the individual organism – its ontogeny – is a separate matter from the evolution of the species to which it belongs. While what an organism does during its life is both a consequence of, and has consequences for, the evolution of its kind, its life history is not part of that evolution. In its neo-Darwinian conception, evolution is *not* a life process. If we ask what evolves, it is not the living organism itself, nor its manifest capabilities of action, but rather a formal design specification for the organism, its genotype. By definition, the genotype is given independently of any particular environmental context of development. Its evolution takes place over numerous generations, through gradual changes brought about by natural selection in the frequency of its information-bearing elements, the genes. Ontogeny is then understood as the process whereby the genotypic specification is translated, within a certain environmental context, into the manifest form of the phenotype.

Most contemporary biologists regard the phenotype as the outcome of an interaction, over the course of a life cycle, between genotype and environment. Indeed this is often called the 'first law of biology'. But the formula is misleading on two counts. First, since 'environment' apparently includes everything relevant to the development of an organism barring the genes themselves, genes cannot interact *with* an environment but only *in* an environment with other entities that are, of course, also interacting with one another. So why should the environment always be defined in relation to the genes rather than to any other of the myriad interactants in the cell? The answer, second, is that the equivalence accorded to genes and environment in the interactionist formula is an illusion. For the distinction between genes and environment is mapped on to a more ancient distinction in Western thought, that between form and substance. Thus the genotype is privileged as the locus of organic form, while the environment is supposed merely to provide the material conditions for its substantive realisation. While an organism may develop different features in changed environments, these differences are regarded as merely alternative phenotypic 'expressions' of the same basic design. Only when the design itself changes is evolution said to occur.

For humankind, it follows that it must be possible to specify what a human being *is*, independently of the manifold conditions of development under which humans live. This possibility is entailed in the assumption that human beings together make up a species – that is, a class of entities that may be grouped together on the grounds of their possession of certain design features transmitted along lines of descent from a common ancestral source. The sum of these features amounts to what many call 'human nature'. This idea was around long before

Darwin. What Darwin added was the claim that human nature is the product of an evolutionary process. So if walking is part of human nature, it must have its basis in a design specification – a programme for the assembly of a functioning bipedal apparatus – that has evolved alongside the rest of the genotypic endowment each of us receives at conception. Thus humans are said to be universally equipped with an innate capacity to walk on two feet, regardless of how they walk in practice, or of whether they walk at all – or go everywhere by car! Specific ways of walking have not themselves evolved, they are just alternative phenotypic realisations of a pre-established, genotypic trait.

2 *Cognitive science.* Just as neo-Darwinian biology presumes a context-independent specification for the design of the body, so cognitive science posits an independent specification for the architecture of the mind. This includes the various cognitive mechanisms or processing devices which would have to be in place before any kind of transmission of cultural representations could take place. Cognitive scientists generally assume that the problem of the origin of these mechanisms has already been solved by evolutionary biology. Since the information specifying the mechanisms cannot be transmitted culturally, there is only one possibility: it must be transmitted genetically. Indeed, by and large in the literature of cognitive science, the postulation of innate mental structures receives no more justification than vague references to genetics and natural selection.

Just as evolution has provided humans with a body that can walk and a vocal apparatus that allows them to speak, so, we are told, it has also furnished a mind with the acquisition devices that enable them to take on representations for walking in culturally particular ways and for speaking particular languages. However, this union of evolutionary biology and cognitive science is not without its contradictions, which are proving to be a particular source of difficulty for the new discipline of evolutionary psychology to which it has given birth. The trouble lies with the distinction between innate and acquired structures (see Pat Bateson's chapter). This distinction lies at the heart of cognitive science's account of how the mind works. A mind without innate mechanisms – that is, one conceived as a 'blank slate' – apparently could not learn, since it would have no way of making sense of the data of experience. And without learning there could be no transmission of representations across generations and hence no culture.

Yet the majority of evolutionary biologists have long since discarded the innate/acquired dichotomy. The architecture of the organism, they say, is neither innate nor acquired, but the outcome of a lifelong interaction between endogenous, genetic factors and exogenous, environmental ones (see e.g. Steven Rose's chapter). It is one thing to claim that every organism starts out with a design specification (the genotype) encoded in the materials of heredity, quite another to claim – as does cognitive science – that every human is born with a preformed mental architecture. For such an architecture, in order to function, would have to exist not merely in the virtual guise of a design, but already 'hard-wired' in the brain. Somehow or other, in order to kick-start the process of cultural transmission, strands of DNA have magically to transform themselves into information-processing modules (see Gabriel Dover's chapter). This is rather like supposing that merely by replicating the design of an aircraft, on the drawing board or computer screen, one is all prepared for take-off. I return below to the attempts of evolutionary psychology to resolve this dilemma.

3 Culture theory. The final component of the trilogy is a certain notion of culture, conceived as a corpus of knowledge or information that can be transmitted across generations independently of its practical application. This notion goes back to a celebrated definition by anthropologist Ward Goodenough, who in 1957 pronounced, 'A society's culture consists of whatever it is one has to know or believe in order to operate in a manner acceptable to its members.'[5] Notice how this definition effectively separates the process by which cultural knowledge is acquired from the way it is expressed in observable behaviour. One obtains the knowledge *in order* to be able to operate or function in the world. The underlying logic of this separation is identical to that which separates genotype from phenotype in biology. Just as the genotype contains a context-independent specification for the design of the organism, so the transmitted cultural information contains a context-independent specification for its behaviour, consisting of what have variously been described as plans, programmes, schemata, representations, recipes, rules and instructions. And where the genotype is said to be 'realised' in the context-specific form of a certain phenotype, through a process of development within an environment, so is culture said to be 'expressed' in the life history of the individual by way of his or her environmentally situated behaviour.

This analogy, in turn, underwrites theories of so-called gene-culture co-evolution, which start from the premise that in human populations two mechanisms of inheritance or information transmission operate in parallel: one genetic, the other cultural.[6] Each of us receives from our predecessors one set of genes and another set of cultural instructions or 'memes'. (See Mary Midgley's chapter.) Together these pull the strings in the development of behaviour. Though recently popularised by Richard Dawkins, the idea that culture consists of particles of heritable information analogous to genes is scarcely novel. In 1956 the anthropologist Clyde Kluckhohn coined the expression 'cultural geno-type' to refer to the pattern of rules and representations underlying manifest behaviour. Numerous similar suggestions have been advanced since then, one of the more recent coming from sociobiologist E. O. Wilson and his collaborator Charles J. Lumsden, who christen the analogue of the gene the 'culturgen', even going so far as to recommend how the term should be pronounced![7]

To sum up the relation between processes of biogenetic and cultural transmission, as generally understood within the framework of the complementarity thesis, consider a human lineage. In each generation bodily anatomy and mental information-processing capacities are built to genetic specifications established through natural selection. And in each generation the mind's capacities are filled with information from which are built the programmes that put the bodily equipment to use in culturally specific ways. Thus the genes provide the necessary instructions for building a workable bipedal apparatus and vocal tract, as well as for the assembly of the walking acquisition device and the LAD. With the aid of these devices, individuals of each generation are able to take on board the rules that enable them to walk, to paraphrase Goodenough, in a manner acceptable to the members of their society, and to speak correctly in the language of their community.

Thus evolutionary biology, cognitive science and culture theory conspire to produce a synthetic account of the living, acting human as a creature of three components, of genotype, mind and culture. More-over, all three approaches – in biology, psychology and anthropology – share one fundamental premise: that the bodily forms, intellectual capacities and behavioural dispositions of humans are specified inde-pendently and in advance of their involvement in practical contexts of environmental activity. Yet in each of the three disciplines the dominant paradigms have come under attack, and for similar reasons. Neo-Darwinism has been criticised for its inability to offer an adequate account of ontogeny, cognitive science for its removal of the mind from human bodily engagement in the world and culture theory for its separation of knowledge from practical application.

By combining these lines of criticism, coming respectively from developmental biology, ecological psychology and the anthropological theory of practice, it should be possible to produce a counter-synthesis much more powerful than the prevailing biopsychocultural orthodoxy.

In this, to which I now turn, the conventional divisions between body, mind and culture would be dissolved, though not as in the more extreme versions of sociobiology or cultural constructionism, by reducing everything to one or other of these terms. Rather, the synthesis offers a unitary focus on the whole organism-person, undergoing a process of growth and development within an environment and contributing through its presence and activity to the development of others.

Developmental Biology

To begin, I return to the analysis of walking. What does it mean to say, as in the conventional account, that I, along with all my fellow humans, possess a capacity to walk? Do we all, likewise, have a capacity to swim, to relax for long periods in a squatting position or to carry things on our heads? I can indeed swim, though plenty of people cannot. Yet like everyone else who has been brought up to sit on chairs, I find that having to squat for any length of time is acutely uncomfortable – though I am told that, with sufficient training, this can be overcome. Along with the great majority of inhabitants of the Western world, however, I am quite incapable of carrying things on my head, at least without manual support.

Are we to conclude, then, that unlike walking carrying things on the head is not innate to humans but culturally acquired? What of the capacity to read and write? Any catalogue of alleged human universals tends to project the image that people of affluent, Western societies have of themselves. Thus where *we*, as privileged members of such societies, can do things that *they* – people of 'other cultures' – cannot, this is typically attributed to the greater development, in ourselves, of universal human capacities. But where they can do things that we cannot, this is put down to the particularity of their cultural tradition. This is to apply just the kind of double standards that have long served to reinforce the modern West's sense of its own superiority over 'the rest', and its sense of history as the progressive fulfilment of its own ethnocentric vision of human potentials. Once we level the playing field of comparison, however, only one alternative remains: that all human beings must have been genotypically endowed, at the dawn of history, with the 'capacity' to do everything that they ever have done in the past, and ever will do in the future – not only to walk, talk, swim and squat but also to read and write, do the pole vault, ride on horseback, drive cars or fly aeroplanes.

Of course, human babies are no more born walking and talking than they are born swimming or squatting: these are bodily skills whose development presupposes an environment that includes already competent care givers, a range of supporting objects and surfaces, and a certain medium or terrain. Given the requisite environmental conditions, these skills are more or less bound to develop; yet in every case their development depends on a process of learning through interaction with other persons and things. This must be true of every bodily skill that humans have ever practised, regardless of its degree of cultural particularity, or the level of social or artefactual scaffolding entailed in its acquisition. Thus the notion of capacity is vacuous unless it refers back to the overall set of conditions that must be present, not only in the individual's genetic constitution but also in the surrounding environment, to make the subsequent development of the characteristic or capability in question a realistic possibility. A negative example clarifies the point. Humans may nowadays be able to fly planes, but despite determined attempts, they cannot fly unaided.

Does this not, then, establish some kind of 'bottom line'? Whatever the environmental conditions, there are certain things that humans potentially can do and others that they definitely cannot. Doubtless a great deal of genetic change would be needed to turn a human into something like a bat – enough to rule out the possibility for our immediate descendants! But while the genetic difference may provide some of the explanation for why humans cannot fly and bats cannot walk, it would be a serious mistake to infer from this that the particular genetic constitution of the bat encodes, within itself, a design for constructing the mechanism of flight, or conversely, that human genes encode a design for building the apparatus of bipedalism. The source of the error lies in the identification of genetic differences with formal traits, a trick which, as Paul Weiss noted long ago, automatically vests genes with exclusive responsibility for organisation and order.[8] It is one thing to claim that without certain genetic modifications having taken place in the lines of descent leading, respectively, to bats and humans, bats could not fly and humans could not walk. But it is quite another to speak of the establishment, in these lineages, of 'genes for flying' or 'genes for walking'.

This is not to deny that every organism starts life with its complement of DNA in the genome. But for the genome to encode any kind of design specification, it would be necessary to suppose that some means exists of 'reading off' this specification from the sequence of DNA base pairs which is independent of any developmental process. No such means has ever been demonstrated. There is only one reading of the genome, and that is the process of ontogeny itself. Hence there can be no design for the organism other than its actual phenotypic form, as it emerges within a particular developmental context. The genotype, conceived as a context-independent design specification, does not exist. It follows that the forms and capacities of human and other organisms are attributable, in the final analysis, not to genetic inheritance but to the generative potentials of the developmental system, that is, the entire system of relations constituted by the presence of the organism, including its genes, in a particular environment. As the philosopher of biology Susan Oyama has pointed out, only within the context of such a system can we possibly say what any gene, or cluster of genes, is 'for'.[9] And so too, in the particular case of human beings, there can be no determination of what a human being is, no human nature, apart from the manifold ways in which humans become, as they live out their lives in diverse communities and environments.

Thus an infant learns to walk in the approved manner of his or her society: it is not as though the latter is somehow added on to a generalised bipedality that has appeared of its own accord, in advance of the infant's entry into the world. Hence there is no such thing as 'bipedal locomotion', as distinct from the various ways in which people actually walk; no pre-programmed 'essence' of the activity that is isolable from the real-time performance of the activity itself.[10] And since ways of walking are properties of neither genes nor 'culture' (conceived as a package of transmissible information), but rather of developmental systems, to account for their evolution we have to understand how such systems are constituted and reconstituted over time. The key to this understanding lies in the recognition that humans, like all other creatures, create through their own actions the environmental conditions both for their own future development and for that of others to which they relate. Thus they figure not as passive 'sites' of evolutionary change but as creative agents, producers as well as products of their own evolution (see Steven Rose's chapter). Far from having been fixed genetically, at some time in the ancestral past, skills such as walking continue to evolve in the very course of our everyday lives.

Clearly, neither orthodox evolutionary biology nor its complement in the field of culture theory is able to offer a coherent account of human development. According to the complementarity thesis, every human being is in part ready-made genetically, in part moulded through the superimposition, upon this preformed substrate, of pre-existing norms and values. Thus in childhood, as anthropologist Walter Goldschmidt puts it, the infant is transformed 'from a purely biological being into a culture-bearing one'.[11] Real humans, however, are not like that. Rather, they grow in an environment furnished by the presence and activities of others. To be sure, as people go through life, they grow out of certain ways of doing things and grow into others. But no one has ever grown out of biology or grown into society or culture. We do not progress, in the course of our lives, from a stage of biological incompletion as 'mere' organisms, to one of social completion as fully fledged persons. We are all fully and indissolubly organism and person from beginning to end.

Walking is certainly biological in that it is part of the way human organisms work. But it is only thanks to the person's involvement in a social world that he or she can undergo normal development as an organic being. A condition for learning to walk is that there is a ground surface to walk on and that this condition is universally fulfilled. Yet how could the infant, taking his or her first steps, encounter 'ground', as a concrete condition of development, not only as distinct from, but also prior to, such diverse 'walk-on-able' surfaces as sand, asphalt, meadow and heath, all of which call for different modalities of gait, balance and footwork? Human infants do not learn to walk in isolation, and even adults rarely walk alone. In everyday practice a person's movements, his or her step, gait and pace, are continually responsive to the movements of others in the immediate environment. Indeed it is largely in this responsiveness that the skill of walking lies. And it is in this respect, too, rather than in its expression of values that somehow reside in an extra-somatic domain of collective representations, that walking is pre-eminently social.

Ecological Psychology

Thus a design specification for the organism cannot be derived from its genetic constitution alone, independently of the conditions of development in an environment. Turning now from evolutionary biology to cognitive science, the problem is further compounded. If the theory of learning as the transmission of cultural information is to work, the requisite processing devices must already exist, not merely – as it were – 'on the drawing board', but in the concrete hard-wiring of human brains. Attempts to resolve this problem, insofar as it is even recognised, are confused and contradictory, and boil down to two distinct claims. One is that the concrete mechanisms making up what evolutionary psychologists call the mind's 'evolved architecture' are reliably constructed, or 'wired up', under all possible circumstances. The other is that these universal mechanisms work on variable inputs from the environment to produce the diversity of manifest capabilities that we actually observe.

Consider the case of language acquisition. Here the alleged universal mechanism is the so-called language acquisition device. All human infants, regardless of the language or languages they may end up speaking in later life, are supposed to come pre-equipped with such a device. The evolutionary psychologists John Tooby and Leda Cosmides (without citing any evidence) claim that even individuals raised in isolation, though they may never learn to speak, will nevertheless possess 'the same species-typical language acquisition device as everyone else'.[12] During a well-defined stage of development, this device is said to be activated, operating on the input of speech sounds from the environment so as to install, in the infant's mind, the grammar and lexicon of the particular language spoken in his or her community. It would thus appear that language acquisition is a two-stage process: in the first, the LAD is constructed; in the second, it is put into service in order to furnish the capacity so established with specific syntactic and semantic content. Notice how this model of cognitive development depends on factoring out those features of the environment that are constant, or reliably present, in every conceivable developmental context, from those that represent a source of variable input from one context to another. Only the former are relevant in the first stage (the construction of innate mechanisms); only the latter are relevant in the second (the acquisition of culturally specific capabilities). The notion that competence in his or her mother tongue is acquired upon the base of a preformed 'language instinct' partitions out the child's experience on the environment just as in my walking example.

Of course, for comparative analytic purposes it is sometimes helpful, even essential, to sift the general from the particular, or to establish a kind of 'lowest common denominator' of development. But real environments are not partitioned in this way. In the case of language development, from well before birth infants are sensitive to the surrounding ambience of sound and above all to the mother's voice. Thus the baby comes into the world already attuned to certain environmentally specific sound patterns. From birth onwards, it is surrounded by an entourage of variously accomplished speakers who provide support in the form both of contextually grounded interpretations of the infant's vocalisations and of demonstrations, or attention-directing gestures, to accompany their own. This environment, then, is not a source of variable input for a preconstructed 'device', but rather furnishes the variable conditions for the development of the neurophysiological structures underwriting the child's capacity to speak. As the conditions vary, so these structures will take manifold forms, each 'tuned' both to specific sound patterns and to other features of local contexts of utterance. These variably attuned structures, and the competencies they establish, correspond to the diverse languages of the world. In short, language – in the sense of the child's capacity to speak in the manner of his or her community – is not acquired. Rather, it is continually being generated and regenerated in the developmental contexts of children's involvement in worlds of speech. And if language is not acquired there can be no such thing as an innate language-learning device.[13]

What applies in the case of language and speech can be extended to other aspects of cultural competence. Thus learning to walk, like learning to talk, is a matter not of acquiring *from* an environment representations that satisfy the input conditions of some preconstituted cognitive device, but of the formation *within* an environment of the necessary anatomy, neurological connections and musculature that underwrite the skill. In short, the systems that actually generate skilled activity are not hard-wired but 'softly assembled'.[14] This conclusion, however, puts paid to one of the key ideas behind the thesis of the complementarity of body, mind and culture, namely that cultural learning is like filling a universal, genetically specified container with culturally specific content. The notion that culture is transmissible from one generation to the next as a corpus of knowledge, independently of its application in the world, is untenable for the simple reason that it rests on the impossible precondition of a ready-made cognitive architecture. The condition is impossible because no matter at what point in the life cycle one might choose to identify a particular structure or mechanism – even before birth – a history of development in a certain environment already lies behind it.

In truth, nothing is really transmitted at all. For the growth of practical knowledge in the life history of a person is a result not of information transmission but of guided rediscovery. In each successive generation, novices learn through being placed in situations in which, faced with certain tasks, they are shown what to do and what to watch out for, under the tutelage of more experienced hands. To show something to someone is to cause it to be made present for that person, so that he or she can apprehend it directly, whether by looking, listening or feeling. Placed in a situation of this kind, the novice is instructed to attend to this or that aspect of what can be seen, touched or heard, so as to get the 'feel' of it for him- or herself. This is not a matter of replicating memes, culturgens or any other such particles of cultural information. For what each generation contributes to the next are not rules and representations for the production of appropriate behaviour, but rather the specific circumstances under which successors, growing up in a social world, can develop their own embodied skills and dispositions, and their powers of awareness and response. Learning in this sense is what the ecological psychologist James Gibson calls an 'education of attention'.[15]

Ecological psychologists reject the view that individuals acquire the knowledge needed to operate in the external world through a processing, in the mind, of sensory inputs delivered to it from the receptor organs of the body. This view artificially separates the activity of the mind in the body from the reactivity of the body in the world, and in so doing merely perpetuates a mind–body split that has bedevilled our thinking since the days of Descartes. An ecological approach, on the contrary, takes as its point of departure the condition of the whole organism-person, indivisibly body and mind, actively engaged with salient components of the environment in the practical tasks of life. Humans, like other animals, get to know the world directly by moving about in the environment and discovering what it affords, rather than by representing it in the mind. Thus meaning, far from being added by the mind to the flux of raw sensory data, is continually being generated within the relational contexts of people's practical engagement with the world around them.

In line with his ecological principles, Gibson maintained that we learn to perceive by a fine-tuning or sensitisation of the entire perceptual system to particular features of our surroundings. Through this process, the human emerges not as a creature whose evolved capacities are filled up with structures that represent the world, but rather as a centre of awareness and agency whose processes resonate with those of the environment. Knowledge, then, far from lying in the relations between structures in the world and structures in the mind, mediated by the person of the knower, is immanent in the life and experience of the knower as it unfolds within the field of practice set up through his or her presence as a being-in-the-world. With this conclusion we have reached the point where we can cross the final barrier, from the psychology of perception to the anthropology of cultural difference.

The Anthropological Theory of Practice

According to the complementarity thesis, culture consists of packages of rules and representations available for transmission across generations, independently of their practical application. These add up to what cognitive anthropologists call 'cultural models'. Culture, suggests Bradd Shore in a recent volume, is best seen as a very large and heterogeneous collection of such models. Corresponding closely to the schemas of psychologists, these models are said to furnish people with 'what they must know in order to act as they do, make the things they make, and interpret their experience in the distinctive way that they do'.[16] But just as an organism's genotype is allegedly unaffected by the vagaries of its life history, so the knowledge held in the cultural models is supposed to remain aloof from the 'hands on' business of doing, making and experiencing. Acquired from predecessors and stored in memory, whence it will be passed along to successors, this knowledge is supposed to be expressed in practice, but not to be generated in it.

However, this view of culture – as what Roy D'Andrade has called ' "pass it along" type information'[17] – has not gone unchallenged. Indeed there is a powerful counter-movement that would reject it altogether. One of the most influential figures in this has been Pierre Bourdieu. In a series of works dedicated to the elaboration of a theory of practice, Bourdieu has attempted to show how knowledge, rather than being imported by the mind into contexts of experience, is itself generated within these contexts in the course of people's involvement with others in the ordinary business of life. He is referring to the kind of practical knowhow that we associate with skill – a knowhow that we carry in our bodies and that is notoriously refractory to codification in terms of rules and representations. Think of the technique involved in tying shoelaces, or breaking an egg, or doing the ironing. Such skills are developed not through formal instruction but through the repeated and often wordless performance of tasks involving specific postures and gestures. Together these furnish a person with his or her orientations in the world.[18]

One could say that such skills and orientations are 'embodied'. This is not to suggest, as some cultural theorists do, that the human body should be understood as a kind of surface upon which social and cultural content can be inscribed. Such a view would render the body passive and reduce its movements to mere signs, directing attention elsewhere in the search for what they stand for: a realm of attitudes, beliefs or mental states that floats like a mirage above the road we tread in real life.[19] The point is rather that in treading this road the body undergoes processes of growth and decay, and concomitantly particular skills, habits, capacities and strengths, as well as debilities and weaknesses, are enfolded into its very constitution – in its neurology, musculature, even its anatomy. To adopt a distinction suggested by the social historian Paul Connerton, this is a matter of incorporation rather than inscription.[20]

Having arrived at this point, however, I can see no further justification for upholding a distinction between the body and the organism. Surely the body – with its powers of autonomous movement, active and alive to the world – *is* the organism. But so, for that matter, is the mind. Indeed, one could just as well speak of 'enmindment' as of 'embodiment', for to develop certain patterns of movement in the world is, at one and the same time, to develop certain modalities of attending to it. If mind, as Gregory Bateson so passionately argued, 'is not limited by the skin', but rather extends outwards into the environment along the multiple pathways of sensory involvement,[21] so likewise the body is not a static, self-contained entity but given in movement, undergoing continual growth and development along the lines of its manifold environmental relationships. Body and mind, therefore, are not two separate things but two ways of describing the same thing – or better, the same process, namely the activity of the organism–person in his or her environment. What, after all, is movement, as Thelen asks, 'but a form of perception, a way of knowing the world as well as acting on it?'[22] Walking, for example, could be described as a way of getting about, but equally as a way of getting to know the environment, primarily by way of contact through the feet, but also thanks to the sights and sounds that the movement affords.

These observations offer a new resolution to one of the oldest of anthropological conundrums. Take two people from different backgrounds, and place them in the same situation: they will differ in what they make of it. Why should this be so? Cognitive anthropologists would respond that it is because they are handling the same input of sensory data in terms of dissimilar cultural models or representational schemas. The theory of practice, however, suggests an alternative answer. Our two characters perceive their surroundings differently because they have been trained, through previous experience of carrying out diverse practical tasks involving particular bodily movements and sensitivities, to orient themselves in relation to the environment and to attend to its features in different ways. The difference, in other words, lies not in the ways in which people represent the environment inside their heads, but in the ways they discover what it affords for their activities. Crucially, this implies that how people perceive will depend upon how they move, including how they walk. I have noted already that a large part of the skill of walking lies in the responsiveness of one's movement to the movements of others in the vicinity. But it responds, too, to ever-changing conditions in the non-human environment. That it does so is immediately apparent if we pause to imagine what walking would be like if it did not.

As an illustration, let me return to Marcel Mauss's observations on the subject of walking. We laugh, he said, at the goose-stepping German soldier. Why? Because his movements are so oddly mechanical. No one naturally walks like that: indeed if they did, they would forever be tripping over things. The goose-step is only possible on the level parade ground. Under ordinary conditions the human gait, though rhythmic, is never metronomic, nor do the feet or knees follow exactly the same trajectory from step to step. Thus walking cannot be reduced to the mechanical application of a fixed motor programme or formula. Nor can any other form of skilled practice. Hence also, it cannot be through the replication of such formulae that skills are passed from generation to generation. Traditional models of social learning, as we have seen, separate the intergenerational transmission of information specifying particular techniques from the application of this information in practice. First, a model or generative schema is said to be established in the novice's mind from observations of the movements of already accomplished practitioners; second, the novice is supposed to imitate these movements by running off exemplars of the technique in question from the schema. Undeniably, the learning of skills involves both observation and imitation. My contention, however, is that the novice's observation of accomplished practitioners is not detached from, but grounded in, his own active, perceptual engagement with his surroundings. And the key to imitation lies in the intimate co-ordination of the movement of the novice's attention to others with his own bodily movement in the world.

This is what is meant, colloquially, by getting the 'feel' of things. And it is this, too, that marks the progression from clumsiness to dexterity. The clumsy practitioner is one who implements mechanically a sequence of received instructions, while remaining insensitive to the evolving conditions of the task as it unfolds. Conversely, to have a feel for what one is doing means moving in a way that is continually and subtly responsive to the nuances of one's relations with relevant aspects of the environment. To achieve such fluency of performance it is not enough to observe; one has also to undertake repeated practical trials. But in these trials it is the task, rather than the precise trajectory of movement, that is repeated. The novice engaged in such trials is not 'acquiring culture', as though it could be simply downloaded into his or her head from a superior source in society, but is rather embarked upon the process that anthropologist Jean Lave has called 'understanding in practice'.[23] And our conclusion, that such understanding calls for a fine

241

tuning of skills of action and perception through repeated trials within an environment, is fully consistent not only with the ecological approach in psychology, reviewed above, but also with a biological focus on the generative dynamics of developmental systems.

Conclusion

At this point we can return to the claim of Clifford Geertz that while all humans come into the world with the 'natural equipment' to live any kind of life, in the end they live only one. Human life, in this view, is conceived as a movement from the universal to the particular, or from biology to culture, entailing a gradual filling up of capacities and a closing down of possibilities. This view, I argue, is fundamentally mistaken. Our bodily equipment is not ready-made but undergoes continual formation in the course of our lives. Even the skeleton, for example, grows in a body that is actively doing things, and its precise form may bear the mark of these activities.[24] Bodily growth, moreover, is an aspect of the very same developmental process by which we gain proficiency in the skills appropriate to the particular kind of life we lead. So what is already in place at the moment of inauguration of a new human life cycle? Not an open-ended design specification in the guise of the genotype, but rather the total system of relations comprised by the presence of the fertilised egg with its complement of DNA, in a womb, in the body of the mother-to-be, who in turn is alive and active within a particular environment. In short, what each of us begins with is a developmental system.

Humans are not born biologically or psychologically identical prior to their differentiation by culture. There has to be something wrong with any explanatory scheme that needs to base itself on the manifestly ludicrous claim – in the words of John Tooby and Leda Cosmides – that 'infants are everywhere the same'.[25] Even parents of identical twins know this to be untrue! The source of the difficulty lies in the notion that culture is an extra ingredient that has to be 'added in' so as to complete the human being. On the contrary, all those specific abilities that have classically been attributed to culture are in reality incorporated, through processes of development, as properties of human organisms. In that sense, they are fully biological. Culture, then, is not super-organic or supra-biological. It is not something added to organisms but a measure of the differences between them. And these differences arise from the ways in which they are positioned vis-à-vis one another, and non-human components of the environment, in wider fields of relationship.

Now if, by evolution, we mean differentiation and change over time in the forms and capacities of organisms, then we must surely allow that skills like walking in a certain way, speaking a certain language, and so on, being biological properties of organisms, must have evolved. We cannot, however, attribute this evolution to changing gene frequencies. No one would seriously suggest that people from different backgrounds walk and talk in different ways merely because of differences in their genetic make-up. But nor does it make sense to suppose that the differences are due to something else, namely culture, which overwrites a generalised biological substrate. Walking and talking are no more the operations of an encultured mind than they are of a body designed by natural selection. They are rather the developmentally enhanced achievements of the whole organism-person, at once body and mind, positioned within an environment. And to account for these achievements we need nothing less than a new approach to evolution, one that sets out to explore not the variation and selection of intergenerationally transmitted attributes, but the self-organising dynamics and form-generating potentials of relational fields.

Of course, cumulative changes may take place, over successive generations within a population, in the frequencies with which particular genes are represented. These changes can be explained, at least in part, by the logic of natural selection. However, what I do deny is the existence of any link between changes in gene frequencies, and in the forms and capacities of organisms, which is independent of the dynamics of development. In orthodox evolutionary biology this link is established by way of the concept of the genotype. Remove this concept and you take away the keystone, without which the entire edifice of neo-Darwinian theory collapses. Natural selection, in short, may occur within evolution, but does not explain it. Only by going beyond the theory of evolution through variation under natural selection, and by considering the properties of dynamic self-organisation of developmental systems, can we hope to discover the possible consequences of those changes that can be explained by natural selection for the evolutionary process itself.

The root source of the explanatory poverty of neo-Darwinian theory is not hard to find. It lies in what one of its principal architects, Ernst Mayr, calls 'population thinking'.[26] Modern biology, Mayr insists, requires us to think of evolutionary change as aggregated over populations of numerous discrete individuals, each of which is uniquely specified in its essential constitution independently of, and prior to, its life in the world. This way of thinking, however, systematically disrupts any attempt to understand the generative dynamics of developmental systems. How can one hope to grasp the continuity of the life process through a mode of thought that can only countenance the organic world already shattered into a myriad of fragments? All it can do is to count up the pieces. What we need, instead, is a quite different way of thinking about organisms and their environments. I call this 'relational thinking'. It means treating the organism not as a discrete, prespecified entity but as a particular locus of growth and development within a continuous field of relationships. It is a field that unfolds in the life activities of organisms and that is enfolded (through processes of embodiment or enmindment) in their specific morphologies, powers of movement and capacities of awareness and response. Our conception of evolution, then, is more topological than statistical. But only with such a conception can we understand the evolutionary process from within, recognising that we ourselves are no more capable of watching from the sidelines than are creatures of any other kind, and that like them, we participate with the whole of our being in the continuum of organic life.

Notes and References

1 Clifford Geertz, *The Interpretation of Cultures* (New York, Basic Books, 1973), p. 45.
2 Marcel Mauss, *Sociology and Psychology: Essays*, trans. B. Brewster (London, Routledge & Kegan Paul, 1979), pp. 102, 114–15.
3 Ibid., pp. 101.
4 Roy G. D'Andrade, 'The Cultural Part of Cognition', *Cognitive Science*, 5 (1981), pp. 179–95.
5 Cited in R. G. D'Andrade, 'Cultural Meaning Systems', in Richard A. Shweder and Robert A. LeVine (ed.), *Culture Theory: Essays on Mind, Self and Emotion* (Cambridge, Cambridge University Press, 1984), p. 89.
6 William H. Durham, *Coevolution: Genes, Culture and Human Diversity* (Stanford, CA, Stanford University Press, 1991).
7 R. W. Gerard, Clyde Kluckhohn and Anatol Rapoport, 'Biological and Cultural Evolution: Some Analogies and Explorations', *Behavioral Science*, 1 (1956), pp. 6–34; Charles J. Lumsden and E. O. Wilson, *Genes, Mind and Culture* (Cambridge, MA, Harvard University Press, 1981), p. 7.
8 Paul Weiss, 'The Living System: Determinism Stratified', in Arthur Koestler and J. R. Smythies (ed.), *Beyond Reductionism: New Perspectives in the Life Sciences* (London, Hutchinson, 1969), p. 35.
9 Susan Oyama, *The Ontogeny of Information: Developmental Systems and Their Evolution* (Cambridge, Cambridge University Press, 1985).
10 This point has been powerfully argued, with specific reference to walking,

by Esther Thelen. See Esther Thelen, 'Motor Development: A New Synthesis', *American Psychologist*, 50 (1995), pp. 79–95.

11 Walter Goldschmidt, 'On the Relationship Between Biology and Anthropology', *Man* (N.S.) 28 (1993), pp. 341–59.

12 John Tooby and Leda Cosmides, 'The Psychological Foundations of Culture', in Jerome H. Barkow, Leda Cosmides and John Tooby (ed.), *The Adapted Mind: Evolutionary Psychology and the Generation of Culture* (New York, Oxford University Press, 1992), p. 45.

13 A. DeCasper and M. Spence, 'Prenatal Maternal Speech Influences Newborns' Perception of Speech Sounds', *Infant Behavior and Development*, 9 (1986), pp. 13–50; Patricia Zukow-Goldring, 'A Social Ecological Realist Approach to the Emergence of the Lexicon: Educating Attention to Amodal Invariants in Gesture and Speech', in Cathy Dent-Read and Patricia Zukow-Goldring (ed.), *Evolving Explanations of Development: Ecological Approaches to Organism-Environment Systems* (Washington, DC, American Psychological Association, 1997); C. H. Dent, 'An Ecological Approach to Language Development: An Alternative Functionalism', *Developmental Psychobiology*, 23 (1990), pp. 679–703.

14 On this idea, see Thelen, 'Motor Development'; and Andy Clark, *Being There: Putting Brain, Body and the World Together Again* (Cambridge, MA., MIT Press, 1997), pp. 42–5.

15 James J. Gibson, *The Ecological Approach to Visual Perception* (Boston, Houghton Mifflin, 1979), p. 254.

16 These lines are quoted from Naomi Quinn and Dorothy Holland, 'Culture and Cognition', in Dorothy Holland and Naomi Quinn (ed.), *Cultural Models in Language and Thought* (Cambridge, Cambridge University Press, 1987), p. 4. See Bradd Shore, *Culture in Mind: Cognition, Culture and the Problem of Meaning* (New York, Oxford University Press, 1996), p. 44.

17 D'Andrade, 'The Cultural Part of Cognition', p. 179.

18 Pierre Bourdieu, *Outline of a Theory of Practice* (Cambridge, Cambridge University Press, 1977), p. 87.

19 On these points, see Michael Jackson, *Paths Toward a Clearing: Radical Empiricism and Ethnographic Inquiry* (Bloomington, Indiana University Press, 1989), pp. 122–3.

20 Paul Connerton, *How Societies Remember* (Cambridge, Cambridge University Press, 1989), pp. 72–3.

21 Gregory Bateson, *Steps to an Ecology of Mind* (London, Granada, 1973), p. 429.

22 Thelen, 'Motor Development', p. 89.

23 Jean Lave, 'The Culture of Acquisition and the Practice of Understanding', in James W. Stigler, Richard A. Shweder and Gilbert Herdt (ed.), *Cultural Psychology: Essays on Comparative Human Development* (Cambridge, Cambridge University Press, 1990), pp. 309–27.

24 See, for example, Theya Molleson, 'The Eloquent Bones of Abu Hureyra', *Scientific American*, 271 (1994), pp. 60–5.

25 Tooby and Cosmides, 'The Psychological Foundations of Culture', p. 33.

26 Ernst Mayr, *The Growth of Biological Thought* (Cambridge, MA, Harvard University Press, 1982), pp. 45–7.

15

Escaping Evolutionary Psychology

Steven Rose

The declared aim of evolutionary psychology is to provide explanations for the patterns of human activity and the forms of organisation of human society which take into account the fact that humans are animals and, like all other currently living organisms, are the present-day products of some four billion years of evolution. So far, so good. The problem with evolutionary psychology is that, like its predecessor, sociobiology, it offers a false unification, pursued with ideological zeal. Far from creating a genuine integration, it offers yet another reductionist account in which presumed biological explanations imperialise and attempt to replace all others. In order to achieve this vain goal evolutionary psychology misspeaks and impoverishes modern biology's understanding of living systems in three key areas: the processes of evolution, of development and of neural function. Underlying all of these are two major conceptual errors: the misunderstanding of the relationship between enabling and causal mechanisms, and the attempt to privilege distal over proximal causes. It is on these shaky foundations that prescriptions for how humans do and must behave, and for the social policies that flow from this, are based. My hope in this chapter is that, by exposing the biological misunderstandings that underlie evolutionary psychology's claims, I will also offer pointers towards the richer concept of living processes towards which I believe the biological sciences should aim.

The Evolutionary Claims of Evolutionary Psychology

There are two distinct strands of thought within the twentieth-century history of biological determinism. The first, which derives from genetics, is the argument that assumes socially relevant *differences* between individuals and groups – for instance, in sexual orientation or intelligence – can be accounted for by genetic differences. The second claims to account for, not differences but assumed *universals* in human nature (such as male aggression or female coyness) by selection pressures derived from human evolutionary past. Evolutionary psychologists go to some lengths to insist that, unlike exponents of earlier versions of Social Darwinism, they are not genetic determinists, or as they sometimes put it, nativists. Rather, they argue, as indeed would most modern biological and social scientists (with the possible exception of psychometricians and behaviour geneticists) that the nature/nurture dichotomy is a fallacious one. Indeed, they are often at pains to distinguish themselves from behaviour geneticists, and there is some hostility between the two.[1] Instead they seek to account for what they believe to be universals in terms of a version of Darwinian theory – a version which in practice owes more to Dawkins's reductive fundamentalism than it does to Darwin's own more pluralistic and observation-rich insights.

In the version of evolutionary theory popularised by Richard Dawkins, and given a philosophical imprimatur by Daniel Dennett, in their books discussed in previous chapters, the fundamental unit of life is a gene, a conceptual abstraction clothed in the biochemistry of the nucleic acid DNA. The purpose or telos of this gene is replication – to make copies of itself – copies which because of random chemical and physical processes may be more or less accurate. The particular chemical structure of DNA provides a mechanism whereby such faithful copying can readily occur – as James Watson and Francis Crick pointed out in the famous finale to their 1953 *Nature* paper describing a proposed molecular structure for DNA.[2] Genes are thus naked replicators and in their struggle to achieve the maximum numbers of identical copies find themselves, mainly in competitive, sometimes in co-operative, relationships with other genes. However, genes cannot achieve replication by themselves; to do so they need to be embedded in cells, which are in organisms. These external manifestations of the work of the genes are formally known as phenotypes, or, to use Dawkins's now famous – or infamous – phrase, are the 'lumbering robots' programmed and set in motion by the genes they contain, whose function is to enable more copies of these genes to be achieved.[3] That is, to reproduce.

Genes within an individual organism share 'an interest' in that organism's successful reproduction and hence may co-operate. However, genes in different individuals within a species are not necessarily identical and hence produce non-identical organisms. These organisms may be more or less 'fit', and hence more or less able to survive to reproduce in their turn. As a result, genes which contribute to the 'fitness' of the lumbering robots in which they are embedded are themselves more likely to survive and spread in the population of such organisms. This variation and spread is the mechanism and process of natural selection. Fitness is, of course, a relative term; it refers only to the specific environment in which the genes and their robot-organisms are located; in different or changing environments different genes and their robots may be advantaged.

This is in essence the 'modern synthesis' of Darwin and Mendel achieved in the 1930s by Ronald Fisher and J. B. S. Haldane. Based on a series of relatively straightforward equations, it also took the study of evolution out of meticulously observed natural history and located it within a more abstract mathematised theory. Indeed, evolution itself came to be defined not in terms of organisms and populations, but as the rate of change of gene frequencies within any given population. One consequence has been a tendency for theoretical evolutionists to retreat further and further into abstract hypotheticals based on computer simulations, and to withdraw from that careful experimentation and patient observation of the natural world which so characterised Darwin's own 'method'.

The additional refinement to the theory, which is essential for the full flowering of evolutionary psychology, was suggested in an often-recounted pub comment by Haldane in the 1950s, that he would be prepared to sacrifice his life for two brothers or eight cousins. Because he shared genes in common with his kin, the proportion varying with the closeness of the kinship, then 'his' genes would serve their interest in replication not only by ensuring that he himself had offspring (which to his regret he did not; he appears to have been infertile), but by aiding in the replication of the identical copies of themselves present in his relatives. This insight was formalised by William Hamilton in the 1960s by redefining fitness as 'inclusive fitness' (i.e. referring to the ability of the genes possessed both by you and your genetic relatives to spread in the population).[4] This is kin selection, and in the 1970s became the core theory of E. O. Wilson's 'new synthesis' of sociobiology, which has, as Hilary Rose describes, mutated into evolutionary psychology. The mathematical syllogism that this version of ultra-Darwinism provides is, within its own framework, irrefutable, which is why Dennett is able to refer to it as a 'universal acid' which eats through all other understandings of not merely biological, but cultural phenomena.[5] Indeed the processes of selection through variation and competition have even been extended to the non-living world; Lee Smolin[6] has claimed that multiple universes – and even perhaps the now no longer 'timeless' 'laws' of physics – have evolved by Darwinian mechanisms.

As Stephen Jay Gould's chapter describes, the central aspect of evolutionary theory upon which evolutionary psychology builds is that of adaptation. Because genes can only survive via the capacity of the organisms they inhabit to survive and replicate, the products of gene action – their phenotypes – must be 'designed' by the honing force of natural selection to achieve that replication with maximal efficiency. Hence every feature of the phenotype, from the protein structures within its cells to its behavioural responses to environmental contingencies, must be considered as adaptations to achieve this goal. In less fit organisms, the adaptations may not work so well, or, because of genetic mutation, may be positively dysfunctional, resulting in disease, incapacity to reproduce, or early death.

Because humans are as subject as any other organism to evolutionary processes, we should therefore expect to find such adaptations amongst our own kind just as much as amongst the others we study. Individual aspects of being human – from our body shape to our eyes and capacity for binocular vision – are clearly evolved features and fit us for the environment in which we live. However, an important aspect of being human, which distinguishes us from other living forms, is our unique social organisation, achieved by virtue of our large brains and the minds they constitute (or at least enable). According to evolutionary psychology, we must consider these too as evolutionary products. The 'architecture' of our minds, and our forms of social organisation, must be seen as adaptations shaped by natural selection to achieve the goals of the optimal replication of the genes of individual humans and the increase in inclusive fitness predicted by the theories of kin selection.

It is here that, despite their disavowals, the nativism of the evolutionary psychology enthusiasts becomes clear. When the anthropologist Laura Betzig writes, as she does in her introduction to her edited collection *Human Nature*,[7] that everything from pregnancy complications to the attraction of money can be explained by invoking Darwinian mechanisms, the conclusion is inescapable: genetic mechanisms underlie these human problems and proclivities. More sophisticated evolutionary psychologists finesse the argument by maintaining that they are not really saying that there are specific actual genes – lengths of DNA – for being attracted sexually by symmetrical bodies, or disliking spinach whilst a child but liking it as an adult, to take but two of the explanatory claims of evolutionary psychology. Instead there are mechanisms, ultimately encoded in the genes, which ensure that we *are*, on average, so attracted, albeit the way in which the genes ensure this is mediated by more proximal mechanisms – by, for instance, creating modular minds whose architecture predisposes this type of behaviour. The spread of such theoretical pseudo-genes, and their putative effect on inclusive fitness can then be modelled satisfactorily *as if* they existed, without the need to touch empirical biological ground at all. Dawkins is particularly prone to this type of stilt-walking, though at least he is open about the as-ifness of the exercise in which he is engaged.[8] For others who join the evolutionary psychology bandwagon not from zoology, like Dawkins, but as psychologists like Steven Pinker or philosophers like Dennett, the disclaimers merely serve as ritual invocations before they get down to the real determinist/nativist business.

Enabling Versus Causing; Proximal Versus Distal Explanations

A consistent feature of this mode of explanation is to mistake enablement for causation. Clearly, all aspects of our existence – and indeed those of any living organism – are made possible by the framing limitations of physical and chemical processes (such as the requirement for energy-providing biochemical reactions, or the structural limitations of calcium phosphate in bone for load-bearing). They are also made possible by the specificities of the organisation of our cells and the multitude of macromolecules of which they are composed. These specificities have themselves been formed during evolution and development. However, this does not entitle one to say, for any observed behaviour, that it is *caused* by such processes; rather, they have made it possible – along with many alternative possibilities. Thus the creation and perception of visual art is dependent on evolved human capacities – having sense organs (eyes) which can respond differentially to wavelengths within a given, relatively limited range, having hands which can wield instruments with which to depict, specific brain structures enabling painting technique and production of pigments to be learned – and a social organisation which makes such production and appreciation possible and valued. None of these capacities is directly causal of the art which is produced, however.

Characteristically, however, evolutionary psychology theorists have argued the reverse. Thus Pinker, in *How the Mind Works*, claims that (with the engaging exception of what he describes as 'great art') humans show a universal propensity to prefer pictures containing green landscapes and water. (See Charles Jencks's chapter for a comment on such 'Bayswater Road' art.) Pinker speculates that this preference may have arisen during human evolution in the African savannah, the so-called Environment of Evolutionary Adaptation, or EEA, discussed by Gould. The grander such assertions, the flimsier and more anecdotal becomes the evidence on which they are based. Has Pinker ever seen savannah, one wonders? Is this so-called universal preference shared by Inuits, Bedouins, Amazonian tribespeople ... Or is it, like so much research in the psychological literature, based on the samples most readily available to American academics – their own undergraduate students? It is hard not to be reminded of the argument once made by an ophthalmologist that El Greco's characteristic elongated figures were the result of his astigmatism.

The point is that there are much simpler proximal explanations for such preferences, should they occur – that in Western urban societies, as Simon Schama points out, 'the countryside' and 'the wilderness' have become associated with particular Arcadian and mythic qualities of escape from life's more pressing problems.[9] Such proximal mechanisms, which relate to human development, history and culture, are much more evidence-based as determining levels of causation, should these be required, than evolutionary speculations. It is surely an essential feature of effective science and of useful explanation to find an appropriate – determining – level for the phenomenon one wishes to discuss. As an example consider the flurry of attempts, particularly in the USA, to 'explain' violent crime by seeking abnormal genes or disordered biochemistry, rather than observing the very different rates of homicide by firearms between, say, the US and Europe, or even within the United States over time, and relating these to the number and availability of handguns.[10] Despite the implicit and sometimes even explicit claim that if we can find an evolutionary explanation for a phenomenon, this will help us fix it,[11] it seems highly doubtful that evolutionary psychology or behaviour genetics will ever contribute anything useful to either art appreciation or crime prevention. The enthusiasm in such cases for proposing biological causal mechanisms owes more to the fervour of the ultra-Darwinians to achieve what E. O. Wilson calls consilience than to serious scholarship.

Evolutionary Time

A further characteristic feature of the evolutionary psychology argument is to point to the relatively short period, in geological and evolutionary terms, over which *Homo sapiens* – and in particular modern society – has appeared. Forms of behaviour or social organisation which evolved adaptively over many generations in human hunter-gatherer society may or may not be adaptive in modern industrial society, but have, it is claimed, become to a degree fixed by humanity's evolutionary experience in the palaeolithic EEA. Hence they are now relatively unmodifiable, even if dysfunctional.

There are two troubles with such claims. The first is that the descriptions that evolutionary psychology offers of what human hunter-gatherer societies were like read little better than Just So accounts, rather like those museum – and cartoon – montages of hunter-dad bringing home the meat whilst gatherer-mum tends the fireplace and kids, so neatly decoded by Haraway in *Primate Visions*.[12] There is a circularity about reading this version of the present into the past, and then claiming that this imagined past explains the present.

However, the more fundamental point is the assertion by evolutionary psychologists that the timescale of human history has been too short for evolutionary selection pressures to have produced significant change. The problem with this is that we know very little about just how fast such change can occur. Evolutionarily modern humans appeared some 100,000 years ago. Allowing 15–20 years as a generation time, there have been some 5,000–6,600 generations between human origins and modern times. Whilst it is possible to calculate mutation rates and hence potential rates of genetic change, such rates do not 'read off' simply into rates of phenotypic change. As Gould and Niles Eldredge have pointed out in developing their theory of punctuated equilibrium, the fossil record shows periods of many millions of years of apparent phenotypic stasis, punctuated by relatively brief periods of rapid change.[13] This is simply because, despite the genetic assumptions of evolutionary psychologists, there is not a one-for-one relationship between gene and organism, or even genotype and phenotype. The many levels of mediation between them means that genetic change can accumulate slowly until at a critical point it becomes canalised into rapid and substantial phenotypic change.

A 'Darwin' is the term used to provide a measure of the rate of evolutionary change. It is based on how the average proportional size of any feature alters over the years and is defined as one unit per million years. Laboratory and field experiments in species varying from fruit flies to guppies give rates of change of up to 50,000 Darwins. Steve Jones describes how English sparrows transported to the south of the USA have lengthened their legs at a rate of around 100,000 Darwins, or 5 per cent a century.[14] So we really have no idea whether the 6,000 or so generations between early and modern humans is 'time enough' for substantial evolutionary change. We don't even know what 'substantial' might mean in this context. However, granted the very rapid changes in human environment, social organisation, technology and mode of production that have clearly occurred over that period, one must assume significant selection pressures operating. It would be interesting in this context to calculate the spread of myopia, which is at least in part heritable, and must in human past have been selected against, once the technological and social developments occurred which have made the almost universal provision of vision-correcting glasses available in industrial societies. What is clear, however, is that the automatic assumption that the palaeolithic was an EEA in which fundamental human traits were fixed, and that there has not been time since to alter them, does not bear serious inspection.

Beyond Ultra-Darwinism

But there are more fundamental problems yet. Even before we get to the human condition, each of the foundational premises of evolutionary psychology is at best only partial, at worst in error. Whilst it is the case, as the population geneticist Theodosius Dobzhansky put it, that nothing in biology makes sense except in the light of evolution, his claim requires extension. Nothing in biology makes sense except in the light of history – the evolutionary history of the species, the developmental history of the individual living organism, and, for humans, of course, social, cultural and technological history. To this must be added the history of our own sciences, which provides the framing assumptions within which we attempt to view and interpret the world. The ontogeny of evolutionary psychology's ways of thinking about the living world – its roots in sociobiology and before that eugenic and Social Darwinist thinking, discussed by Hilary Rose and Ted Benton – goes a long way towards explaining both its current agenda and its biological misconceptions.

To summarise the central inadequacies of the biological theorising on which evolutionary psychology is based:

Naked replicators are empty abstractions. One of the central features of DNA as a molecule is that it cannot simply and unaided make copies of itself; it cannot therefore 'replicate' in the sense that this term is usually understood. It is a relatively inert molecule (hence the possibility of preservation in amber and the plot of *Jurassic Park*). What brings DNA to life, so to speak, is the cell in which it is embedded. Replication – using one strand of the double helix of DNA to provide the template on which another can be constructed – requires an appropriately protected environment, the presence of a wide variety of complex molecular precursors, a set of protein enzymes and a supply of chemical energy. All these are provided in the complex metabolic web within which the myriad biochemical and biophysical interactions occurring in each individual cell are stabilised.[15] Despite the constantly recurring metaphors of DNA as a 'master molecule', a 'blueprint' and so forth, there are no such master molecules in cellular processes. Even the metaphor of the cellular orchestra, which I have used previously, is not adequate, as orchestras require conductors. Better to see cells as marvellously complex versions of string quartets or jazz groups, whose harmonies arise in a self-organised way through mutual interactions. This is why the answer to the chicken and egg question in the origin of life is not that life began with DNA and RNA, but that it must have begun with primitive cells which provided the environment within which nucleic acids could be synthesised and serve as copying templates.[16]

The relationship between genes and phenotypes is not linear. Neither cells nor organisms – still less their behaviours – spring fully formed from DNA, even with the richer account of its synthesis and replication described above. There isn't even a direct relationship between a strand of DNA ('a gene') and a particular protein, which is the immediate gene product. Most of the DNA in the human genome plays no known functional role in the survival and reproduction of the organism, but gets copied nonetheless. Hence it is described disparagingly either as 'junk DNA' or, in the continuation by Crick of the metaphor made notorious by Dawkins, as 'selfish'. As Gabriel Dover emphasises, the modern picture which biochemistry provides is of complex processes of so-called 'editing', 'splicing' and otherwise working on the original DNA sequence before the fully formed protein is generated. And the pattern of these editing processes is itself environmentally shaped. To put this in the language of genetics rather than development, and to use Dobzhansky's terminology, how genes are expressed reflects a *norm of reaction* to the environment.

Beyond the cellular level, there is the little matter of development, the processes that transform the single fused cell of a fertilised ovum into the hundred trillion cells of the human body, hierarchically and functionally organised into tissues and organs. Developmental processes have trajectories which constitute the individual lifeline of any organism, trajectories which are neither instructed by the genes, nor selected by the environment, but constructed by the organism (a process described by Humberto Maturana and Francisco Varela as autopoiesis) out of the raw materials provided by both genes and environment.[17]

This process of self-construction is nicely illustrated in the context of instinct in general, and language acquisition in particular, by Pat Bateson and Annette Karmiloff-Smith. Here I would want to generalise it to embrace the entire process of ontogeny, cellular and physiological as well as behavioural.

The unity of an organism is a process unity, not a structural one. All its molecules, and virtually all its cells, are continuously being transformed in a cycle of life and death which goes on from the moment of conception until the final death of the organism as a whole. This means that living systems are open. They are never in thermodynamic equilibrium. And they are not mere passive vehicles, sandwiched between the demands of their genes and the challenges of their environments. Rather, organisms actively engage in constructing their environments, constantly choosing, absorbing and transforming the world around them. Every living creature is in constant flux, always at the same time both *being* and *becoming*. To build on an example I owe to Pat Bateson, a newborn infant has a suckling reflex; within a matter of months the developing infant begins to chew her food. Chewing is not simply a modified form of suckling, but involves different sets of muscles and physiological mechanisms. The paradox of development is that a baby has to be at the same time a competent suckler and to transform herself into a competent chewer. To be, therefore, and to become.

Being and becoming cannot be partitioned into that tired dichotomy of nature versus nurture. Rather they are defined by a different dichotomy, that of specificity and plasticity. Consider the problem of seeing and of making sense of the world we observe, processes subserved by eye and brain. The retina of the eye is connected via a series of neural staging posts to the visual cortex at the back of the brain. A baby is born with most of these connections in place, but during the first years of life the eye and the brain both grow, at different rates. This means that the connections between eye and brain have continually to be broken and remade. If the developing child is to be able to retain a coherent visual perception of the world this breaking and remaking must be orderly and relatively resistant to modification by the exigencies of development. This is specificity. However, as both laboratory animal experiments and our own human experience show, both the fine details of the 'wiring' of the visual cortex, and how and what we perceive of the world are directly and subtly shaped by early experience. This is plasticity. All living organisms and perhaps especially humans with our large brains show both specificity and plasticity in development, and both properties are enabled by our genes and shaped by our experience and developmental contingency. Neither genes nor environment are in this sense determinant of normal development; they are the raw materials out of which we construct ourselves.

Thus the four dimensions of living processes – three of space and one of time – cannot be read off from the one-dimensional strand of DNA – nor yet the modular minds it is supposed to generate. A living organism is an active player in its own destiny, not a lumbering robot responding to genetic imperatives whilst passively waiting to discover whether it has passed what Darwin described as the continuous scrutiny of natural selection.

Individual genes are not the only level of selection. Central to the theoretical structure of evolutionary psychology is the concept of the individualistic, 'selfish gene' as not only *the* replicator, but as *the only* level at which selection operates. It was this assumption which lay at the core of the Fisher–Haldane synthesis and which envisaged genes rather as individual beads on a string – a view disrespectfully described as the 'bean-bag' model by other population geneticists. The present-day understanding of the fluid genome (see Gould's and Dover's chapters), in which segments of DNA responsible for coding for subsections of proteins, or for regulating these gene functions, are distributed across many regions of the chromosomes in which the DNA is embedded, and are not fixed in any one location but may be mobile, makes this simple view untenable. But it always was. To play their part in the creation of a functioning organism many genes are involved – in the human, some hundred thousand. For the organism to survive and replicate, the genes are required to work in concert – that is, to co-operate. Antelopes that can outrun lions are more likely to survive and breed than those that cannot. Therefore a mutation in a gene which improves muscle efficacy, for instance, might be regarded as fitter and therefore likely to spread in the population. However, as enhanced muscle use requires other physiological adaptations – such as increased blood flow to the muscles, without this concerted change in other genes, the individual mutation is scarcely likely to prove very advantageous. And as many genes have multiple phenotypic effects (pleiotropy) the likelihood of a unidirectional phenotypic change is complex – increased muscle efficacy might diminish the longevity of the heart, for example. Thus it is not just single genes which get selected, but also genomes. Selection operates as the level of gene, genome and organism.

Nor does it stop there. Organisms exist in populations (groups, demes). Back in the 1960s, V. C. Wynne-Edwards argued that selection occurred at the level of the group as well as the individual. He based this claim on a study of a breeding population of red grouse on Scottish moors and argued that they distributed themselves across the moor, and regulated their breeding practices, in a way which was optimum for the group as a whole rather than any individual member within it. It may be in the individual's interest to produce lots of offspring; but this might overcrowd the moor, which could only sustain a smaller number of birds; hence it is in the *group's* interest that none of its members over-breed. Orthodox Darwinians, led by George Williams, treated this claim with as much derision as they did Lamarck's view that acquired characteristics could be inherited, and group selection disappeared from the literature for three decades.

Today, however, it is clear that the attack was misjudged. The work of one of the leaders of current evolutionary orthodoxy, John Maynard Smith, itself indicates that stability in a population of social animals may require the mutual interactions of members with very different types of behaviours – so-called evolutionary stable strategies. Maynard Smith gives as an example a population in which some individuals are aggressive towards others in the group, whereas others are more pacific – he calls them 'hawks' and 'doves'. He goes on to show mathematically that populations with all hawks or all doves are unstable, whereas a mix of both behavioural types at appropriate ratios will be stable.[18] An evolutionary stable strategy like this can be modelled as if it was based on either individual or group selection; the distinction is more semantic than 'real'. But there are an increasing number of examples of populations of organisms whose behaviours can most economically be described by group selectionist equations. Recently Elliott Sober and David Sloan Wilson[19] have published a major reassessment of such models and shown mathematically how even such famously counter-intuitive (for ultra-Darwinians) phenomena as altruism can occur, in which an individual sacrifices its own individual fitness, not merely for the inclusive fitness of its kin but for the benefit of the group as a whole.

Finally, there is selection at even higher levels – that of the species, for example.[20] Natural selection may be constantly scrutinising and honing the adaptiveness of a particular species to its environment, but cannot predict the consequences of dramatic changes in that environment, as for example the meteor crash into the Yucatan believed to have precipitated the demise of the dinosaurs. Selection also operates at the level of entire ecosystems. Consider, for example, a beaver dam. Dawkins uses this example to claim that the dam may be regarded as part of the beaver's phenotype – thus swallowing an entire small universe into the single strand of DNA.[21] But if it is a phenotype, it is the phenotype of many beavers working in concert, and indeed of the many commensal and symbiotic organisms which also live on and modify for their needs the structure of the dam. As Sober and Wilson point out, selection may indeed occur at the level of the individual, but what constitutes an individual is very much a matter of definition. Genes are distributed across genomes within an organism, and they are also distributed across groups of organisms within a population. There is no overriding reason why we should consider 'the organism' as an individual rather than 'the group' or even 'the ecosystem'.

Natural selection is not the only mode of evolutionary change. Darwin was a pluralist. As Benton and Gould repeatedly point out, he was very careful to state that natural selection is not the only motor of evolutionary change. He invented the concept of sexual selection (with all the problems to which Anne Fausto-Sterling refers), the only addition to natural selection which evolutionary psychology theorists are prepared to include in their pantheon. We need not be Lamarckian to accept that other processes are at work. The existence of neutral mutations, founder effects, genetic drift, Gould's exaptations and Dover's adoptions all enrich the picture.

Not all phenotypic characters are adaptive. A core assumption of ultra-Darwinism is that if not all then most observed characters must be adaptive, so as to provide the phenotypic material upon which natural selection can act. However, what constitutes a character – and what constitutes an adaptation – is as much in the eye of the beholder as in the organism that is beheld. Bateson discusses Thayer's suggestion that the flamingo's pink coloration is an adaptation to make them less visible to predators against the pink evening sky. But the coloration is a consequence of the flamingo's shrimp diet and fades if the diet changes. Thus even if we were to assume that the coloration was indeed protective, it is an epiphenomenal consequence of a physiological – dietary – adaptation, rather than a selected property in its own right.

Natural selection's continual scrutiny does not give it an *à la carte* freedom to accept or reject genotypic or phenotypic variation. Structural constraints insist that evolutionary, genetic mechanisms are not infinitely flexible but must work within the limits of what is physically or chemically possible (for instance, the limits to the size of a single cell occasioned by the physics of diffusion processes, the size of a crustacean like a lobster or crab by the constraints of its exoskeleton, or the impossibility of genetically engineering humans to sprout wings and fly because of the limits to the possible lift of any conceivable wing structure.) Webster and Goodwin[22] have extended this argument further, arguing that there are exact 'laws of form' that ensure, for instance pentadactyl (five-fingered) limbs. Gould[23] takes a contrasting line, arguing that much evolutionary change is contingent, accidental, and that, as he puts it, if one were to wind the tape of history backward and replay it, it is in the highest degree unlikely that mammals, let alone humans, would evolve. It is not necessary to adjudicate between these positions to appreciate the limits to the automatic assumption that all phenotypic characters are the honed consequences of natural selection at work.

And yet this is exactly what evolutionary psychology theorists do again and again. Consider Martin Daly and Margo Wilson's search for evolutionary explanations to support their claim that stepfathers kill their adopted offspring more frequently than do natural fathers, ignoring the much more obvious proximal causal processes discussed by Hilary Rose. And such examples manifest themselves repeatedly in the assemblage of papers on human sociobiology and evolutionary psychology that now fill the pages of the evolutionary psychology group's house journals, web sites and academic or popular books.

'Architectural' Minds

Unlike earlier generations of genetic determinists, evolutionary psychologists argue that these proximal processes are not so much the direct product of gene action, but of the evolutionary sculpting of the human mind. The argument, drawing heavily on the jargon and conceptual framework of artificial intelligence, goes as follows: the mind is a cognitive machine, an information-processing device instantiated in the brain. But it is not a general-purpose computer; rather it is composed of a number of specific modules (for instance, a speech module, a number sense module,[24] a face-recognition module, a cheat-detector module, and so forth). These modules have, it is argued, evolved quasi-independently during the evolution of early humanity, and have persisted unmodified throughout historical time, underlying the proximal mechanisms that traditional psychology describes in terms of motivation, drive, attention and so forth. Whether such modules are more than theoretical entities is unclear, at least to most neuroscientists. Indeed evolutionary psychologists such as Pinker go to some lengths to make it clear that the 'mental modules' they invent do not, or at least do not necessarily, map on to specific brain structures. (In this sense they are rather like Dawkins's theoretical genes.) But as Karmiloff-Smith shows, even if mental modules do exist they can as well be acquired as innate.

And even one of the founders of modular theory, Jerry Fodor, has taken some pains to dissociate himself from his seeming followers, in a highly critical review of Pinker.[25]

For evolutionary psychology, minds are thus merely surrogate mechanisms by which naked replicators enhance their fitness. Brains and minds have evolved for a single purpose, sex, as the neuroscientist Michael Gazzaniga rather bluntly puts it.[26] And yet in practice evolutionary psychology theorists, who, as Hilary Rose points out, are not themselves neuroscientists or even, by and large, biologists, show as great a disdain for relating their theoretical constructs to material brains as did the now discredited behaviourist psychologists they so despise. Although many brain processes – such as visual analysis, for example – do take place in distinct 'modules' or cell assemblies,[27] the coherent result of brain activity unifies these distinct processes through distributed, non-hierarchical mechanisms.[28]

The insistence of evolutionary psychology theorists on modularity puts a particular strain on their otherwise heaven-made alliance with behaviour geneticists. For instance, IQ theorists, such as the psychometrician Robert Plomin, are committed to the view that intelligence, far from being modular, can be reduced to a single underlying factor, g, or 'crystallised intelligence'.[29] A similar position has emphatically been taken in recent years by Herrnstein and Murray in *The Bell Curve*,[30] who argue that whatever intelligence is, it cannot be dissociated into modules!

Modules or no, it is not adequate to reduce the mind/brain to nothing more than a cognitive, 'architectural' information-processing machine. Brains/minds do not just deal with information. They are concerned with living meaning.[31] In *How the Mind Works* Pinker offers the example of a footprint as conveying information. My response is to think of Robinson Crusoe on his island, finding a footprint in the sand. First he has to interpret the mark in the sand as that of a foot, and recognise that it is not his own. But what does it mean to him? Pleasure at the prospect of at last another human being to talk and interact with? Fear that this human may be dangerous? Memories of the social life of which he has been deprived for many years? A turmoil of thoughts and emotions within which the visual information conveyed by the footprint is embedded. The key here is emotion, for the key feature which distinguishes brains/minds from computers is their/our capacity to experience emotion. Indeed, emotion is primary – which may be why Darwin devoted an entire book to it rather than to cognition.

Emotional mechanisms and indeed their expression are evolved properties, and several neuroscientists have devoted considerable attention to the mechanisms and survival advantages of emotion.[32] So it is therefore all the more surprising to find this conspicuous gap in the concerns of evolutionary psychologists – but perhaps this is because not even they can speak of a 'module' for emotion. Rather, affect and cognition are inextricably engaged in all brain and mind processes, creating meaning out of information – just one more reason why brains aren't computers. What is particularly egregious in this context is the phrase, repeated frequently by Leda Cosmides, John Tooby and their followers, 'the architecture of the mind'. Architecture, which implies static structure, built to blueprints and thereafter stable, could not be a more inappropriate way to view the fluid dynamic processes whereby our minds/brain develop and create order out of the blooming buzzing confusion of the world which confronts us moment by moment.

On Free Will

There is an ultimate contradiction at the core of evolutionary psychology theory. Whatever the claimed evolutionary honing of our every intention and act, evolutionary psychologists remain anxious to insist on at least their own autonomy. 'If my genes don't like it,' says Pinker, 'they can go jump in the lake.'[33] Rather less demotically, Dawkins insists that only we as humans have the power to rebel against the tyranny of the selfish replicators.[34] Such a claim to a Cartesian separation of these authors' minds from their biological constitution and inheritance seems surprising and incompatible with their claimed materialism. Where does this strange free will come from in a genetically and evolutionarily determined universe?

The problem is indicated even sharply by the reprinting of a series of classical anthropological and sociobiological papers in the collection entitled *Human Nature*.[35] The editor's view is that these show the way that Darwinian insights transform our understanding of social organisation. The papers were largely published in the 1970s and 1980s, and, for their republication in 1997, each author was asked to reflect in retrospect on their findings. What is interesting is that when the anthropologists go back to their field subjects, they report rapid changes in their styles of living. Kipsigis women no longer prefer wealthy men (Borgerhoff, Mulder), the Yanonomo are no longer as violent as in the past (Chagnon), wealth no longer predicts the number of children reared (Gaulin and Boster) and so forth. Each of these societies has undergone rapid economic, technological and social change in the last decade. What has happened to the evolutionary psychology predictions?

Why have these assumed human universals suddenly failed to operate? Has there been a sudden increase in mutation rates? Have the peoples they had studied taken Dawkins to heart and decided to rebel against the tyranny of their selfish replicators?

There is a simpler explanation. The evolutionary path that leads to humans has produced organisms with profoundly plastic, adaptable brains/minds and ways of living. Humans have created societies, invented technologies and cultures. We, the inheritors of not merely the genes, but also the cultures and technologies of our forebears, are profoundly shaped by them in ways that make our future as individuals, societies and species radically unpredictable. In short, the biological nature of being human enables us to create individual lives and collective societies whose futures lie at least in part in our own hands.

Notes and References

1 See, for example, Matt Ridley, *Genome: The Autobiography of a Species in 23 Chapters* (London, Fourth Estate, 1999).

2 'It has not escaped our notice that the specific pairing we have postulated immediately suggests a possible copying mechanism for the genetic material.' J. D. Watson and F. H. C. Crick, 'Molecular Structure of Nucleic Acids', *Nature*, 171 (1953), pp. 737–8.

3 Richard Dawkins, *The Selfish Gene* (Oxford, Oxford University Press, 1976).

4 William D. Hamilton, 'The Genetical Evolution of Social Behaviour, I and II', *Journal of Theoretical Biology*, 7 (1964), pp. 1–32.

5 Daniel Dennett, *Darwin's Dangerous Idea: Evolution and the Meanings of Life* (Harmondsworth, Allen Lane, 1995).

6 Lee Smolin, *The Life of the Cosmos* (London, Weidenfeld & Nicolson, 1997).

7 Laura Betzig (ed.), *Human Nature: A Critical Reader* (New York, Oxford University Press, 1997).

8 See, for instance, his discussion in the first two chapters of *The Extended Phenotype: The Gene as the Unit of Selection* (Oxford, W. H. Freeman, 1982).

9 Simon Schama, *Landscape and Memory* (London, HarperCollins, 1995).

10 For a discussion of the US Violence Initiative see, e.g., Peter R. Breggin and Ginger Ross Breggin, *A Biomedical Program for Urban Violence Control in the US: The Dangers of Psychiatric Social Control* (Washington, Center for the Study of Psychiatry [mimeo], 1994).

11 See Betzig, *Human Nature*.

12 Donna Haraway, *Primate Visions: Gender, Race and Nature in the World of Modern Science* (London, Routledge, 1989).

13 Niles Eldredge, *Time Frames* (New York, Simon & Schuster, 1985).

14 Steve Jones, *Almost Like a Whale* (London, Doubleday, 1999), p. 242.

15 Stuart Kauffman, *At Home in the Universe: The Search for Laws of Complexity* (London, Viking, 1995).

16 For a fuller discussion of this view of the origin of life, see my *Lifelines: Biology Freedom, Determinism* (Harmondsworth, Penguin, 1998).

17 *Lifelines*; Susan Oyama, *The Ontogeny of Information* (Cambridge, Cambridge University Press, 1985); Humberto R. Maturana and Francisco J. Varela, *The Tree of Knowledge: The Biological Roots of Human Understanding* (Boston, Shambhala, 1998).

18 John Maynard Smith, *Evolution and the Theory of Games* (Cambridge, Cambridge University Press, 1982).

19 Elliott Sober and David Sloan Wilson, *Unto Others: The Evolution and Psychology of Unselfish Behavior* (Cambridge, MA, Harvard University Press, 1998).

20 Stephen Jay Gould, 'Gulliver's Further Travels: The Necessity and Difficulty of a Hierarchical Theory of Selection', *Philosophical Transactions of the Royal Society*, Series B, 353 (1998), pp. 307–14.

21 Dawkins, *The Extended Phenotype*.

22 Gerry Webster and Brian Goodwin, *Form and Transformation: Generative and Relational Principles in Biology* (Cambridge, Cambridge University Press, 1996).

23 Stephen Jay Gould, *Wonderful Life; The Burgess Shale and the Nature of History* (London, Hutchinson Radius, 1989).

24 For example see Stanislas Dehaene, *The Number Sense* (New York, Oxford University Press, 1997).

25 Jerry Fodor, *London Review of Books* (January 1998).

26 Michael Gazzaniga, *The Social Brain* (New York, Basic Books, 1985).

27 Semir Zeki, *A Vision of the Brain* (Oxford, Blackwell, 1993).

28 See, e.g., Wolf Singer in Steven Rose (ed.), *Brains to Consciousness? Essays on the New Sciences of the Mind* (Harmondsworth, Allen Lane, 1998), pp. 228–45.

29 Robert Plomin, *British Journal of Psychiatry* (in press); Robert Plomin, John De Fries, Gerald E. McClearn and Michael Rutter, *Behavioral Genetics* (San Francisco, Freeman, 1997).

30 Richard J. Herrnstein and Charles Murray, *The Bell Curve: Intelligence and Class Structure in American Life* (New York, The Free Press, 1994).

31 See, e.g., Steven Rose, *The Making of Memory* (London, Bantam, 1992); Rose (ed.), *Brains to Consciousness*; Walter Freeman, *How Brains Make Up Their Minds* (London, Weidenfeld & Nicolson, 1999).

32 Antonio Damasio, *Descartes' Error: Emotion, Reason and the Human Brain* (New York, Grosset/Putnam, 1994) and *The Feeling of What Happens* (London, Heinemann, 1999); Joseph LeDoux: *The Emotional Brain: The Mysterious Underpinnings of Emotional Life* (New York, Simon & Schuster, 1996).

33 Steven Pinker, *How The Mind Works* (Harmondsworth, Allen Lane, 1997), p. 52.
34 See the concluding sentences of *The Selfish Gene*.
35 Betzig, *Human Nature*.

Notes on Contributors

Patrick Bateson is Professor of Ethology at Cambridge University and was Director of the Sub-department of Animal Behaviour before becoming Provost of King's College, Cambridge, in 1988. His experimental and theoretical interests have centred on the development of behaviour and the role of evolutionary theory in understanding developmental processes, although he is better known to the media in Britain as the author of a study of the behavioural and physiological effects of hunting red deer. His authored and edited books include *Growing Points in Ethology*, *The Domestic Cat* and the series *Perspectives in Ethology*. His most recent book is *Design for a Life* (with Paul Martin).

Ted Benton is Professor of Sociology at the University of Essex. He is author of numerous publications on the philosophy of social science, history of the life sciences and social theory. In recent years he has been exploring ways of linking feminist, green and socialist approaches to social theory and to politics. He is also an amateur naturalist and photographer.

Gabriel Dover is Professor of Genetics at the University of Leicester. An evolutionary geneticist who studies the molecular evolutionary forces which have shaped our genomes and genes, he is the originator of the theory of molecular drive as one of the motors of evolutionary change. Author of many research papers and reviews, his first book for a general readership, *Dear Mr Darwin: Letters on the Evolution of Life and Human Nature*, has recently been published.

Mark Erickson is a sociologist. He has researched and taught at the universities of Durham, Sunderland and Birmingham. His main interests lie in the sociology of science and the sociology of work.

Anne Fausto-Sterling is Professor of Medical Science and Women's

Studies at Brown University, Rhode Island. Her early research focused on the developmental genetics of fruit fly (*Drosophila*) embryos, but her interests have broadened to include the field of gender and science specifically, and science studies, more generally. She is the author of *Myths of Gender: Biological Theories about Women and Men* and is currently working on a new book, tentatively entitled *Body Building: How Biologists Construct Sexuality*.

Stephen Jay Gould, whose books have won many prizes, teaches biology, geology and the history of science at Harvard University. His books include *Ontogeny and Phylogeny*, *Ever Since Darwin*, *Wonderful Life*, *Leonardo's Mountain of Clams and the Diet of Worms* and, most recently, *The Lying Stones of Marrakech*.

Barbara Herrnstein Smith is a professor at Duke University and Director of the Center for Interdisciplinary Studies in Science and Cultural Theory. She has written and edited a number of books on language, literature and critical theory, including *Contingencies of Value: Alternative Perspectives for Critical Theory*, and more recently *Mathematics, Science and Postclassical Theory* and *Belief and Resistance: Dynamics of Contemporary Intellectual Controversy*, which concerns current debates over truth, reason, objectivity and scientific knowledge.

Tim Ingold is a social anthropologist and Professor at Aberdeen University. His fieldwork has been mainly amongst the Saami people of Finland. His current research interests are in the anthropology of technology and skilled practice and in issues of environmental perception. His many authored and edited books include *Evolution and Social Life*, *The Appropriation of Nature*, *Tools, Language and Cognition in Human Nature* and *Key Debates in Anthropology*.

Charles Jencks is an architect, critic and writer, and one of the founders of the Post-Modern Movement. Amongst other appointments, he is Professor at London's Architectural Association. His many books include *What is Post-Modernism?* First written in 1985, it is now in its fourth edition.

Annette Karmiloff-Smith started life as an interpreter for the UN before studying psychology. Her varied career includes working in a Palestinian refugee camp and as a researcher with Jean Piaget in Geneva. She is currently Professor of Psychology at University College London, based at the Institute of Child Health, where she researches infant and child development, especially language acquisition. Her authored and co-authored books include *Beyond Modularity: A Developmental Perspective on Cognitive Science* and *Rethinking Innateness: A*

Connectionist Perspective on Development. She has recently finished a book written with her daughter: *Everything Your Baby Would Ask if Only He/She Could Talk*.

Mary Midgley is a philosopher whose special interests are in the relations of humans to the rest of nature, the sources of morality and the relation between science and religion. Since her retirement from teaching at Newcastle University she has written many books analysing the moral and philosophical claims of the life sciences. They include *Beast and Man*, *Wickedness*, *Science as Salvation* and *The Ethical Primate*.

Dorothy Nelkin, University Professor at New York University, is a sociologist. Her research focuses on controversial areas of science, technology and medicine as a way to understand their social and political implications and the process of decision-making in complex technical areas. She is the author of more than twenty books, including *The Creation Controversy*, *Selling Science* and *The DNA Mystique*.

Tom Shakespeare is a sociologist. He has taught and researched at the universities of Cambridge, Sunderland and Leeds. He currently works for the Policy, Ethics and Life Sciences Research Institute, Newcastle. He is active in the disability movement and has written and broadcast widely on disability and on the new genetics.

Portrack Seminars

The Portrack Seminars are an intermittent set of meetings occurring mostly at Portrack House near Dumfries in Scotland. They have focused on the subject of developments in post-modern culture and the problems that arise from the modern economy and global culture. Brought together by Charles Jencks and his late wife Maggie Keswick, the core members of the group have been Charlene Spretnack, David Ray Griffin and Richard Falk. The group has agreed on a loose agenda of what they term 'Restructive Post-Modernism', a new world-view which, opposed to deconstruction, emphasises the creative potentials which have emerged from the break-up of modern cultures. So far there have been eleven seminars, and this book is largely based on discussion at the tenth meeting.

1 *Post-Modern Culture*, Portrack, 13–17 July 1992. Walter Anderson, Jim Collins, Thomas Docherty, Charles Jencks, Maggie Keswick, Margaret Rose, Charlene Spretnak.
2 *Post-Modern and Science*, Portrack, 7–10 September 1992. Paul Davies, Richard Falk, David Griffin, Mae-Wan Ho, Charles Jencks, Catherine Keller, Maggie Keswick, Ian Marshall, Peter Saunders, Charlene Spretnak, Danah Zohar.
3 *Metaphor and Grounding*, organised by Charles Spretnak, Santa Monica, March 1993. Gordon Globus, David Ray Griffin, Linda Holler, Charles Jencks, Mark Johnson, George Lakoff, Maggie Keswick, Michael Zimmerman.
4 *Post-Modern Cosmology*, ICA, London, 5 July 1993. Bernard Carr, Matthew Fox, Brian Goodwin, Charles Jencks, Mary Midgley, Arthur Peacocke, Charlene Spretnak, Brian Swimme.
5 *Political Foundations of World Order*, organised by David Ray Griffin, Santa Monica, 25–7 February 1994. Robert Benson, John

Cobb, Richard Falk, Robert Hamerton-Kelly, Charles Jencks, Maggie Keswick, Nikki Keddie, George Khutzsisvili, Charlene Spretnak.

6 *The Economic Foundations of World Order*, organised by David Ray Griffin, Portrack, 26–30 June 1994. John Cobb, Paul Ekins, Richard Falk, Robert Hamerton-Kelly, David Held, Charles Jencks, Maggie Keswick, Mary Kaldor, Sara Parkin, Charlene Spretnak.

7 *Post-Modern Ecology*, organised by Michael Zimmerman, Santa Monica, 10–12 March 1995. Chet Bowers, Richard Falk, William Grassie, Charles Jencks, Kevin Kelly, Maggie Keswick, Michael Soule, Charles Spretnak, Gregory Stock.

8 *Intercivilisation – Confrontation and Synergies*, organised by Richard Falk, 3–6 November 1995. Frederick Apffel-Marglin, Amerita Basu, Victoria Bomberry, Hilal Elver, Richard Falk, David Ray Griffin, Charles Jencks, Radha Kumar, Maivan Clech Lam, Marcus Raskin, Lester Edwin J. Ruiz, Sulak Siveraksa, Charlene Spretnak, Kay Warren.

9 *Reconsidering the Post-Modern*, organised by Charles Jencks and Hans Bertens at the ICA, London, 18–19 May 1996. Akbar Ahmed, Hans Bertens, Richard Falk, Anthony Giddens, David Ray Griffin, Ihab Hassan, Andreas Huyssen, Charles Jencks, Charlene Spretnak, Patricia Waugh.

10 *Against Evolutionary Psychology*, organised by Steven Rose and Hilary Rose, Portrack, 25–8 September 1998. Jeremy Ahouse, Patrick Bateson, Ted Benton, Bob Berwick, Gabby Dover, Barbara Herrnstein Smith, Tim Ingold, Charles Jencks, Annette Karmiloff-Smith, Mary Midgley, Dorothy Porter, Tom Shakespeare.

11 *The Architecture of the New Scotland*, organised by Deyan Sudjic and Charles Jencks, Portrack, 12–14 March 1999. Gordon Benson, Angus Kerr, Henry McKeown, Isi Metzstein, Enrick Miralles, Richard Murphy, David Page.

Select Bibliography

Alderson, Priscilla, and Goodey, Chris, *Enabling Education* (London, Tufnell Press, 1998)

Armstrong, A. C., *Transitional Eras in Thought, with Special Reference to the Present Age* (New York, Macmillan, 1904)

Aronson, L. R., Tobach, E., Lehrman, D. S., and Rosenblatt, J. S. (ed.), *Development and Evolution of Behaviour* (San Francisco, Freeman, 1970)

Baker, Robin, *Sperm Wars: The Science of Sex* (New York, Basic Books, 1996)

Barkow, Jerome, Cosmides, Leda, and Tooby, John (ed.), *The Adapted Mind: Evolutionary Psychology and the Generation of Culture* (New York, Oxford University Press, 1992)

Barash, David, *The Whispering Within* (New York, Harper & Row, 1979)

Barrett, Michele, and Phillips, Ann, *Destabilizing Theory* (Cambridge, Polity, 1992)

Bateson, Gregory, *Steps to an Ecology of Mind* (London, Granada, 1973)

Bateson, Patrick (ed.), *The Development and Integration of Behaviour* (Cambridge, Cambridge University Press, 1991)

Bateson, Patrick, and Martin, Paul, *Design for a Life: How Behaviour Develops* (London, Jonathan Cape, 1999)

Benton, Arthur, et al., *Benton Test for Facial Recognition* (New York, Oxford University Press, 1983)

Betzig, Laura (ed.), *Human Nature: A Critical Reader* (New York, Oxford University Press, 1997)

Bishop, Dorothy, *Uncommon Understanding* (Hove, Psychology Press/Erlbaum, 1997)

271

Blackmore, Susan, *The Meme Machine* (Oxford, Oxford University Press, 1999)

Blake, A., and Yuille, A. (ed.), *Active Vision* (Cambridge, MA, MIT Press, 1992)

Boakes, Robert, *From Darwin to Behaviourism* (Cambridge, Cambridge University Press, 1984)

Bourdieu, Pierre, *Outline of a Theory of Practice* (Cambridge, Cambridge University Press, 1977)

Bradley, Harriet, *Fractured Identities* (Cambridge, Polity, 1996)

Breggin, Peter R., and Breggin, Ginger Ross, *A Biomedical Program for Urban Violence Control in the US: The Dangers of Psychiatric Social Control* (Washington, DC, Center for the Study of Psychiatry, mimeo, 1994)

Brimelow, Peter, *Alien Nation* (New York, Random House, 1994)

Broman, S. H., and Grafman, J. (ed.), *Cognitive Deficits in Developmental Disorders: Implications for Brain Function* (Hillsdale, NJ, Erlbaum, 1994)

Brown, Andrew, *The Darwin Wars: How Stupid Genes Became Selfish Gods* (London, Simon & Schuster, 1999)

Brown Blackwell, Antoinette, *The Sexes Throughout Nature* (1875; reprinted Westport, Conn., Hyperion Press, 1976)

Bruce, V., and Young, A., *In the Eye of the Beholder* (Oxford, Oxford University Press, 1998)

Buss, David, *Evolution of Desire: Strategies of Human Mating* (New York, Basic Books, 1994)

Butler, Judith, *Gender Trouble* (New York, Routledge, 1990)

Campbell, Jane, and Oliver, Michael, *Disability Politics* (London, Routledge, 1996)

Campbell, Robin, et al., *Testing Face Processing Skills in Children* (Stirling, Stirling University, 1995)

Carson, R., and Rothstein, M. (ed.), *Behavioral Genetics* (Baltimore, John Hopkins Press, 1999)

Chomsky, Noam, *Rules and Representations* (New York, Columbia University Press, 1980)

Clark, Andy, *Associative Engines: Connectionism, Concepts and Representational Change* (Cambridge, MA, MIT Press, 1993)

Clark, Andy, *Being There: Putting Brain, Body and the World Together Again* (Cambridge, MA, MIT Press, 1997)

Clutton-Brock, T. H., Guinness, F. E., and Albon, S. E., *Red Deer: The Behavior and Ecology of Two Sexes* (Chicago, Chicago University Press, 1982)

Connerton, Paul, *How Societies Remember* (Cambridge, Cambridge University Press, 1989)

Corker, Marian, and French, Sally (ed.), *Disability Discourse* (Buckingham, Open University Press, 1999)

Crook, Paul, *Darwinism, War and History* (Cambridge, Cambridge University Press, 1994)

Daly, Martin, and Wilson, Margo, *Homicide* (New York, Aldine de Gruyter Hawthorne, 1988)

Daly, Martin, and Wilson, Mary, *The Truth about Cinderella: A Darwinian View of Parental Love* (London, Weidenfeld & Nicolson, 1998)

Damasio, Antonio, *Descartes' Error: Emotion, Reason and the Human Brain* (New York, Grosset/Putnam, 1994)

Damasio, Antonio, *The Feeling of What Happens* (London, Heinemann, 1999)

Darwin, Charles, *The Descent of Man* (London, John Murray, 1871)

Darwin, Charles, *The Origin of the Species by Means of Natural Selection* (London, John Murray, 1876)

Darwin, Charles, *The Autobiography of Charles Darwin, 1809–1882. With Original Omissions Restored*, ed. Nora Barlow (New York, Harcourt, Brace, 1958)

Davies, Nicholas B., *Dunnock Behavior and Social Evolution* (New York, Oxford University Press, 1992)

Dawkins, Richard, *The Selfish Gene* (New York and Oxford, Oxford University Press, 1976)

Dawkins, Richard, *The Extended Phenotype: The Gene as the Unit of Selection* (Oxford, W. H. Freeman, 1982)

Dawkins, Richard, *The Blind Watchmaker* (Harlow, Longman, 1986)

Dawkins, Richard, *Climbing Mount Improbable* (Harmondsworth, Penguin, 1997)

Dawson, G., Fischer, K., et al. (ed.), *Human Behavior and the Developing Brain* (New York, Guilford, 1994)

Deacon, Terrence W., *The Symbolic Species: The Co-evolution of Language and the Brain* (New York and London, Norton, 1997)

Dehaene, Stanislas, *The Number Sense* (Oxford, Oxford University Press, 1997)

Dent-Read, Cathy, and Zukow-Goldring, Patricia (ed.), *Evolving Explanations of Development: Ecological Approaches to Organism-Environment Systems* (Washington, DC, American Psychological Association, 1997)

Dennett, Daniel, *Darwin's Dangerous Idea: Evolution and the Meanings of Life* (New York, Simon & Schuster, 1995; Harmondsworth, Penguin, 1996)

Desmond, Adrian, and Moore, James, *Darwin* (Harmondsworth, Penguin, 1992)

Dickens, Peter, *Social Darwinism* (Milton Keynes, Open University Press, 1999)

Douglas, Mary, *Purity and Danger* (Harmondsworth, Penguin, 1976)

Dover, Gabriel, *Dear Mr Darwin: Letters on the Evolution of Life and Human Nature* (London, Weidenfeld & Nicolson, 2000)

Durham, William H., *Co-evolution: Genes, Culture and Human Diversity* (Stanford, CA, Stanford University Press, 1991)

Durkheim, Emile, *Suicide* (1897; London, Routledge & Kegan Paul, 1952)

Edelman, Gerald, *Bright Air, Brilliant Fire: On the Matter of the Mind* (New York, Basic Books, 1992)

Eldredge, Niles, *Time Frames* (New York, Simon & Schuster, 1985)

Eldredge, Niles, *Reinventing Darwin: The Great Debate at the High Table of Evolutionary Theory* (New York, John Wiley, 1995)

Ellman, Jeffrey, Bates, Elizabeth A., Johnson, Mark H., Karmiloff-Smith, Annette, Parisi, Domenico, and Plunkett, Kim, *Rethinking Innateness: A Developmental Perspective on Connectionism* (Cambridge, MA, MIT Press, 1996)

Fausto-Sterling, Anne, *Myths of Gender: Biological Theories About Women and Men* (New York, Basic Books, 1992)

Fodor, Jerry, *The Modularity of Mind* (Cambridge, MA, MIT Press, 1984)

Fukuyama, Francis, *The End of Social Order* (London, Social Market Foundation, 1998)

Foucault, Michel, *The History of Sexuality* (Harmondsworth, Penguin, 1976)

Foucault, Michel, *Politics, Philosophy, Culture* (London, Routledge, 1990)

Fox, Nicholas, *Postmodernism, Sociology and Health* (Buckingham, Open University Press, 1993)

Freeman, Walter, *How Brains Make Up Their Minds* (London, Weidenfeld & Nicolson, 1999)

Gasman, Daniel, *The Scientific Origins of National Socialism* (London, Macdonald, 1971)

Gazzaniga, Michael, *The Social Brain* (New York, Basic Books, 1985)

Geertz, Clifford, *The Interpretation of Cultures* (New York, Basic Books, 1973)

van Gelder, Tim, and Port, Robert F. (ed.), *Mind as Motion: Explorations in the Dynamics of Cognition* (Cambridge, MA, MIT Press, 1995)

Gibson, James J., *The Ecological Approach to Visual Perception* (Boston, Houghton Mifflin, 1979)

Gibson, R., and Ingold, T., *Tools, Language and Cognition in Human Evolution* (Cambridge, Cambridge University Press, 1993)

Gilroy, Paul, *The Black Atlantic: Modernity and Double Consciousness* (London, Verso, 1993)

Goodall, Jane, *The Chimpanzees of Gombe: Patterns of Behavior* (Cambridge, MA, Harvard University Press, 1986)

Gottlieb, G., *Development of Species Identification in Birds* (Chicago, Chicago University Press, 1971)

Gould, Stephen Jay, *The Mismeasure of Man* (New York, Norton, 1981)

Gould, Stephen Jay, *Wonderful Life: The Burgess Shale and the Nature of History* (London, Hutchinson Radius, 1989)

Gould, Stephen Jay, *Lifes Grandeur* (London, Jonathan Cape, 1996; published in the USA as *Full House*, New York, Crown, 1996)

Griffiths Paul E., *What Emotions Really Are: The Problem of Psychological Categories* (Chicago, Chicago University Press, 1997)

Gross, Paul, and Levitt, Norman, *Higher Superstition: The Academic Left and Its Quarrels with Science* (Baltimore, John Hopkins University Press, 1994)

Haeckel, Ernst, *The Pedigree of Man and Other Essays* (London, Freethought, 1883)

Hager, Lori (ed.), *Women in Human Evolution* (London and New York, Routledge, 1997)

Haraway, Donna, *Primate Visions: Gender, Race and Nature in the World of Modern Science* (London and New York, Routledge, 1989)

Helier, T. C., et al. (ed.), *Reconstructing Individualism* (Stanford, Stanford University Press, 1986)

Hendriks-Jansen, Horst, *Catching Ourselves in the Act: Situated Activity, Interactive Emergence, Evolution and Human Thought* (Cambridge, MA, MIT Press, 1996)

Herrnstein, Richard, and Murray, Charles, *The Bell Curve: Intelligence and Class Structure in American Life* (New York, The Free Press, 1994)

Herrnstein Smith, Barbara, *Belief and Resistance: Dynamics of Con-

temporary Intellectual Controversy (Cambridge, MA, Harvard University Press, 1997)

Hirschfeld, L. A., and Gelman, S. A. (ed.), *Mapping the Mind: Domain Specificity in Cognition and Culture* (Cambridge, Cambridge University Press, 1994)

Hogan, J. A., and Bolhuis, J. J. (ed.), *Causal Mechanisms of Behavioural Development* (Cambridge, Cambridge University Press, 1994)

Holland, Dorothy, and Quinn, Naomi (ed.), *Cultural Models in Language and Thought* (Cambridge, Cambridge University Press, 1987)

Horgan, John, *The Undiscovered Mind: How the Brain Defies Explanation* (London, Weidenfeld & Nicolson, 1999)

Hrdy, Sarah Blaffer, *The Woman That Never Evolved* (Cambridge, MA, Harvard University Press, 1981)

Hrdy, Sarah Blaffer, *Mother Nature: Natural Selection and the Female of the Species* (London, Chatto & Windus, 1999)

Hubbard, R., Hennifin, M. S., and Fried, B., *Women Look at Biology Looking at Women* (Cambridge, MA, Schenckman, 1979)

Hutchins, Edwin, *Cognition in the Wild* (Cambridge, MA, MIT Press, 1996)

Jackson, Michael, *Paths Towards a Clearing: Radical Empiricism and Ethnographic Inquiry* (Bloomington, Indiana University Press, 1989)

Jencks, Charles, *The Architecture of the Jumping Universe: How Complexity Science is Changing Architecture and Culture* (London, Academy, 1995 and 1997)

Johnson, Mark, *Developmental Cognitive Neuroscience: An Introduction* (Oxford, Blackwell, 1997)

Jones, Steve, *Almost Like a Whale* (London, Doubleday, 1999)

Karmiloff-Smith, Annette, *Beyond Modularity: A Developmental Perspective on Cognitive Science* (Cambridge, MA, MIT Press, 1992)

Kauffman, Stuart, *At Home in the Universe: The Search for Laws of Complexity* (London, Viking, 1995)

Kimura, M., *The Neutral Theory of Molecular Evolution* (Cambridge, Cambridge University Press, 1983)

Kinzey, W. G. (ed.), *The Evolution of Human Behavior* (Albany, NY, SUNY, 1987)

Koestler, Arthur, and Smythies, J. R. (ed.), *Beyond Reductionism: New Perspectives in the Life Sciences* (London, Hutchinson, 1969)

Kohn, D. (ed.), *The Darwinian Heritage* (Princeton, NJ, Princeton

University Press, 1985)

Kohn, Marek, *The Race Gallery: The Return of Racial Science* (London, Jonathan Cape, 1995)

Kuo, Z. Y., *The Dynamics of Behavioral Development* (New York, Random House, 1967)

Lakoff, George, and Johnson, Mark, *Metaphors We Live By* (Chicago, Chicago University Press, 1980)

Lakoff, George, *Women, Fire and Dangerous Things: What Categories Reveal About the Human Mind* (Chicago, Chicago University Press, 1987)

Latour, Bruno, *We Have Never Been Modern* (Cambridge, MA, Harvard University Press, 1993)

LeDoux, Joseph, *The Emotional Brain: The Mysterious Underpinnings of Emotional Life* (New York, Simon & Schuster, 1996)

Leibowitz, Lila, *Females, Males, Families: A Biosocial Approach* (North Scituate, Duxbury Press, 1978)

Lewin, Roger, *Complexity: Life at the Edge of Chaos* (New York, Macmillan, 1992)

Lieberman, Philip, *Eve Spoke: Human Language and Human Evolution* (New York, Norton, 1998)

Lorenz, Konrad, *Evolution and Modification of Behaviour* (Chicago, Chicago University Press, 1965)

Lumsden, Charles J., and Wilson, E. O., *Genes, Mind and Culture* (Cambridge, MA, Harvard University Press, 1981)

Lyons, John (ed.), *New Horizons in Linguistics* (Harmondsworth, Penguin, 1970)

Macrae, D. G., *Ideology and Society* (London, Heinemann, 1961)

Martin, Emily, *The Woman in the Body: A Cultural Analysis of Reproduction* (Boston, Beacon, 1987)

Marx, Karl, and Engels, Friedrich, *Selected Correspondence 1846–1895* (London, Lawrence & Wishart, 1943)

Maturana, Humberto R., and Varela, Francisco J., *Autopoiesis and Cognition: The Realization of the Living* (Boston, D. Reidel, 1988)

Maturana, Humberto R., and Varela, Francisco J., *The Tree of Knowledge: The Biological Roots of Human Understanding* (Boston and London, Shambala, 1988)

Mauss, Marcel, *Sociology and Psychology: Essays* (London, Routledge & Kegan Paul, 1979)

Mayr, Ernst, *The Growth of Biological Thought* (Cambridge, MA, Harvard University Press, 1982)

Morris, Jenny (ed.), *Encounters with Strangers* (London, Women's Press, 1996)

Nelkin, Dorothy, and Lindee, M. Susan, *The DNA Mystique: The Gene as Cultural Icon* (New York, W. H. Freeman, 1995)

Nisbett, A., *Konrad Lorenz* (London, Dent, 1976)

Noble, David, *The Religion of Technology: The Divinity of Man and the Spirit of Invention* (New York, Knopf, 1997)

Oakley, Ann, *Sex, Gender and Society* (London, Temple Smith, 1972)

Oyama, Susan, *The Ontogeny of Information: Developmental Systems and Evolution* (Cambridge, Cambridge University Press, 1985)

Oyama, Susan, *Evolutions Eye* (Durham, NC, Duke University Press, forthcoming)

Pinker, Steven, *The Language Instinct: How the Mind Creates Language* (New York, William Morrow, 1994)

Pinker, Steven, *How the Mind Works* (New York, Norton, and London, Allen Lane, 1997)

Plomin, Robert, De Fries, John, McClearn, Gerald E., and Rutter, Michael, *Behavioral Genetics* (San Francisco, Freeman, 1997)

Raff, R., *The Shape of Life: Genes, Development and the Evolution of Animal Form* (Chicago, Chicago University Press, 1996)

Rice, M. (ed.), *Towards a Genetics of Language* (Mahway, NJ, Lawrence Erlaum 1996)

Ridley, Matt, *The Origins of Virtue* (London, Viking, 1996)

Ridley, Matt, *Genome: The Autobiography of a Species in 23 Chapters* (London, Fourth Estate, 1999)

Riley, Denise, *Am I That Name?* (London, Macmillan, 1988)

Rose, Nikolas, *Governing the Soul* (London, Routledge, 1989)

Rose, Steven, Lewontin, Richard, and Kamin, Leo, *Not in Our Genes* (Harmondsworth, Penguin, 1984)

Rose, Steven, *The Making of Memory* (London, Bantam, 1992)

Rose, Steven, *Lifelines: Biology, Freedom, Determinism* (Harmondsworth, Allen Lane, 1997)

Rose, Steven (ed.), *From Brains to Consciousness? Essays on the New Sciences of the Mind* (Harmondsworth, Allen Lane, 1998)

Rosenblatt, J. S. (ed.), *Advances in the Study of Behavior* (New York, Academic Press, 1996)

Ross, Andrew (ed.). *Science Wars* (Durham, Duke University Press,

1996)

Runciman, W. G., *The Social Animal* (London, HarperCollins, 1998)

Rushton, J. Philippe, *Race, Evolution and Behavior* (New Brunswick, Transaction Books, 1994)

Sahlins, Marshall, *The Uses and Abuses of Biology* (London, Tavistock, 1977)

Sayers, Janet, *Biological Politics* (London, Tavistock, 1982)

Schama, Simon, *Landscape and Memory* (London, HarperCollins, 1995)

Shore, Bradd, *Culture in Mind: Cognition, Culture and the Problem of Meaning* (New York, Oxford University Press, 1996)

Shweder, Richard A., and LeVine, Robert A. (ed.), *Culture Theory: Essays on Mind, Self and Emotion* (Cambridge, Cambridge University Press, 1984)

Singer, Peter, *A Darwinian Left: Politics Evolution and Co-operation* (London, Weidenfeld & Nicolson, 1999)

Small, Meredith (ed.), *Female Choices: The Sexual Behavior of Female Primates* (New York, Cornell University Press, 1984)

Small, Meredith (ed.), *Female Primates: Studies by Women Primatologists* (New York, A. R. Liss, 1984)

Smith, John Maynard, *Evolution and the Theory of Games* (Cambridge, Cambridge University Press, 1982)

Smith, Neil, and Tsimpli, Ianthi, *The Mind of a Savant: Language Learning and Modularity* (Oxford, Blackwell, 1995)

Smolin, Lee, *The Life of the Cosmos* (London, Weidenfeld & Nicolson, 1997)

Sober, Elliott, and Wilson, David Sloan, *Unto Others: The Evolution and Psychology of Unselfish Behaviour* (Cambridge, MA, Harvard University Press, 1998)

Stigler, James W., Shweder, Richard A., and Herdt, Gilbert (ed.), *Cultural Psychology: Essays on Comparative Human Development* (Cambridge, Cambridge University Press, 1990)

Stringer, Chris, and McKie, Robin, *African Exodus* (London, Jonathan Cape, 1996)

Strum, Shirley, *Almost Human: A Journey into the World of Baboons* (New York, Random House, 1987)

Suchman, Lucy, *Plans and Situated Actions* (Cambridge, Cambridge University Press, 1987)

Symons, Donald, *The Evolution of Human Sexuality* (Oxford and New York, Oxford University Press, 1979)

Tanner, Nancy Makepeace, *On Becoming Human* (Cambridge, Cambridge University Press, 1981)

Tavris, Carol, *The Mismeasure of Women* (New York, Simon & Schuster, 1992)

Thele, Esther, and Smith, Linda B., *A Dynamic Systems Approach to the Development of Cognition and Action* (Cambridge, MA, MIT Press, 1994)

Thornhill, Randy, and Palmer, Craig, *A Natural History of Rape: Biological Bases for Sexual Coercion* (Cambridge, MA, MIT Press, 2000)

Varela, Francisco J., Thompson, Evan, and Rosch, Eleanor, *The Embodied Mind: Cognitive Science and Human Experience* (Cambridge, MA, MIT Press, 1991)

de Vore, Irven, *Primate Behavior: Field Studies of Monkeys and Apes* (New York, Holt, Rinehart & Winston, 1965)

de Waal, Frans, *Peacemaking Among Primates* (Cambridge, MA, Harvard University Press, 1989)

Waldrop, Mitchell, *Complexity: The Emerging Science at the Edge of Order and Chaos* (New York, Simon & Schuster, 1992)

Wallace, Alfred Russel, *The Malay Archipelago* (1869; New York, Dover, 1962)

Wallace, Alfred Russel, *Social Environment and Moral Progress* (London, Cassell, 1913)

Webster, Gerry, and Goodwin, Brian, *Form and Transformation: Generative and Relational Principles in Biology* (Cambridge, Cambridge University Press, 1996)

Weiner, Jonathan, *The Beak of the Finch* (London, Vintage, 1994)

Wellcome Trust Medicine and Society Programme, *Public Perceptions on Human Cloning* (London, Wellcome Trust, 1998)

Wilson, Barbara, et al., *The Rivermead Behavioural Memory Test* (Fareham Thames Valley Test Company, 1985)

Wilson, B. (ed.), *Rationality* (Oxford, Blackwell, 1974)

Wilson, E. O., *Sociobiology: The New Synthesis* (Cambridge, MA, Harvard University Press, 1975)

Wilson, E. O., *On Human Nature* (Cambridge, MA, Harvard University Press, 1978)

Wilson, E. O., *Consilience: The United Knowledge* (New York, Knopf, and London, Little Brown, 1998)

Winch, Peter, *The Idea of a Social Science* (London, Routledge & Kegan Paul, 1958)

Select Bibliography

Wright, Robert, *The Moral Animal: Why We Are the Way We Are. The New Science of Evolutionary Psychology* (New York, Pantheon, 1994)

Zeki, Semir, *A Vision of the Brain* (Oxford, Blackwell, 1993)

Index

adaptation, 2, 10, 409, 86, 88, 89, 90, 91–3, 95, 96, 98, 99–100, 101–5, 114, 118, 121, 122, 123, 130, 135, 170, 178, 217, 218, 250, 259–60

The Adapted Mind (ed. Barkow, Cosmides, Tooby), 129, 130, 139, 219, 221

adoptation, concept of, 10, 63, 64

Alderson, Priscilla and Goodey, Chris, *Enabling Education*, 199–200

algorithms, 93, 96, 97, 105

altruism, phenomenon of, 48

American Association for the Advancement of Science, 14, 25

anthropological theory of practice, 231, 238–42

anthropology, 119, 120, 123, 227, 231, 262; cognitive, 227, 239, 240

Ardrey, Robert, 6; *The Territorial Imperative*, 114

Arecchi, Tito, 35–6

Aristotle, 5, 29

artificial intelligence (AI), 135, 260

Artificial Intelligence Laboratory, MIT, 135

AS (Asperger's Syndrome), 200, 203

Ashworth, John, 113, 114, 115

atomism, 74, 99

Attention Deficit Hyperactivity Disorder, 198, 199

Austen, Jane, 107

autism, 148, 198, 199–200, 201

Baker, Robin, 21; *The Sperm Wars*, 17

Barash, David, 116

Barkow, Jerome H., 130, 139, 185

Bateson, Gregory, 240

Bateson, Patrick, 8, 11, 12, 59, 157–73, 229, 256, 259, 266

Bateson, Patrick and Martin, Paul, *Design for a Life*, 171

Baur, Erwin, 5

Beach, Frank, 163

beauty, female, 39–41, 43

Beauvoir, Simone de, 33

Beck, Ulrich, 119

Beethoven, Ludwig van, 85

Behrens, C. B. A., 69–70

Benton, Ted, 11–12, 109, 175, 182, 190, 206–24, 254, 259, 266

Betzig, Laura, 118; *Human Nature* (ed.), 250, 262

Beveridge, William, 112

biodiversity, 37, 42

biological determinism, 4–6, 8, 211, 216, 218, 219, 247–8; social constructionism and disability, 190–205 *passim*